Solving Nonlinear Equations with Newton's Method

Fundamentals of Algorithms

Editor-in-Chief: Nicholas J. Higham, University of Manchester

The SIAM series on Fundamentals of Algorithms publishes monographs on state-of-the-art numerical methods to provide the reader with sufficient knowledge to choose the appropriate method for a given application and to aid the reader in understanding the limitations of each method. The monographs focus on numerical methods and algorithms to solve specific classes of problems and are written for researchers, practitioners, and students.

The goal of the series is to produce a collection of short books written by experts on numerical methods that include an explanation of each method and a summary of theoretical background. What distinguishes a book in this series is its emphasis on explaining how to best choose a method, algorithm, or software program to solve a specific type of problem and its descriptions of when a given algorithm or method succeeds or fails.

Kelley, C. T.   *Solving Nonlinear Equations with Newton's Method*

C. T. Kelley

North Carolina State University
Raleigh, North Carolina

Solving Nonlinear Equations with Newton's Method

Society for Industrial and Applied Mathematics
Philadelphia

10 9 8 7 6 5 4 3 2

Library of Congress Cataloging-in-Publication Data

Kelley, C. T.
   Solving nonlinear equations with Newton's method / C.T. Kelley.
      p. cm. — (Fundamentals of algorithms)
   Includes bibliographical references and index.
   ISBN-10: 0-89871-546-6 (pbk.)
   ISBN-13: 978-0-898715-46-0 (pbk.)
1. Newton-Raphson method. 2. Iterative methods (Mathematics) 3. Nonlinear theories.
I. Title. II. Series.

QA297.8.K455 2003
511'.4—dc21   2003050663

To my students

Contents

Preface

This small book on Newton's method is a user-oriented guide to algorithms and implementation. Its purpose is to show, via algorithms in pseudocode, in MATLAB®, and with several examples, how one can choose an appropriate Newton-type method for a given problem and write an efficient solver or apply one written by others.

This book is intended to complement my larger book [42], which focuses on indepth treatment of convergence theory, but does not discuss the details of solving particular problems, implementation in any particular language, or evaluating a solver for a given problem.

The computational examples in this book were done with MATLAB v6.5 on an Apple Macintosh G4 and a SONY VAIO. The MATLAB codes for the solvers and all the examples accompany this book. MATLAB is an excellent environment for prototyping and testing and for moderate-sized production work. I have used the three main solvers `nsold.m`, `nsoli.m`, and `brsola.m` from the collection of MATLAB codes in my own research. The codes were designed for production work on small- to medium-scale problems having at most a few thousand unknowns. Large-scale problems are best done in a compiled language with a high-quality public domain code.

We assume that the reader has a good understanding of elementary numerical analysis at the level of [4] and of numerical linear algebra at the level of [23, 76]. Because the examples are so closely coupled to the text, this book cannot be understood without a working knowledge of MATLAB. There are many introductory books on MATLAB. Either of [71] and [37] would be a good place to start.

Parts of this book are based on research supported by the National Science Foundation and the Army Research Office, most recently by grants DMS-0070641, DMS-0112542, DMS-0209695, DAAD19-02-1-0111, and DAAD19-02-1-0391. Any opinions, findings, and conclusions or recommendations expressed in this material are those of the author and do not necessarily reflect the views of the National Science Foundation or the Army Research Office.

Many of my students, colleagues, and friends helped with this project. I'm particularly grateful to these stellar rootfinders for their direct and indirect assistance and inspiration: Don Alfonso, Charlie Berger, Paul Boggs, Peter Brown, Steve Campbell, Todd Coffey, Hong-Liang Cui, Steve Davis, John Dennis, Matthew Farthing, Dan Finkel, Tom Fogwell, Jörg Gablonsky, Jackie Hallberg, Russ Harmon, Jan Hesthaven, Nick Higham, Alan Hindmarsh, Jeff Holland, Stacy Howington, Mac Hyman, Ilse Ipsen, Lea Jenkins, Katie Kavanagh, Vickie Kearn, Chris Kees, Carl and Betty Kelley, David Keyes, Dana Knoll, Tammy Kolda, Matthew Lasater, Debbie Lockhart, Carl Meyer, Casey Miller, Tom Mullikin, Stephen Nash, Chung-Wei Ng, Jim Ortega, Jong-Shi Pang, Mike Pernice, Monte Pettitt, Linda Petzold, Greg Racine, Jill Reese, Ekkehard Sachs, Joe Schmidt, Bobby Schnabel, Chuck Siewert, Linda Thiel, Homer Walker, Carol Woodward, Dwight Woolard, Sam Young, Peiji Zhao, and every student who ever took my nonlinear equations course.

C. T. Kelley
Raleigh, North Carolina
May 2003

How to Get the Software

This book is tightly coupled to a suite of MATLAB codes.

The codes are available from SIAM at the URL

```
http://www.siam.org/books/fa01
```

The software is organized into the following five directories. You should put the SOLVERS directory in your MATLAB path.

(1) **SOLVERS**

- nsold.m   Newton's method, direct factorization of Jacobians
- nsoli.m   Newton–Krylov methods, no matrix storage
- brsol.m   Broyden's method, no matrix storage

(2) Chapter1: solvers for scalar equations with examples

(3) Chapter2: examples that use nsold.m

(4) Chapter3: examples that use nsoli.m

(5) Chapter4: examples that use brsol.m

One can obtain MATLAB from

The MathWorks, Inc.
3 Apple Hill Drive
Natick, MA 01760-2098
(508) 647-7000
Fax: (508) 647-7001
Email: info@mathworks.com
WWW: http://www.mathworks.com

Chapter 1

Introduction

1.1 What Is the Problem?

Nonlinear equations are solved as part of almost all simulations of physical processes. Physical models that are expressed as nonlinear partial differential equations, for example, become large systems of nonlinear equations when discretized. Authors of simulation codes must either use a nonlinear solver as a tool or write one from scratch. The purpose of this book is to show these authors what technology is available, sketch the implementation, and warn of the problems. We do this via algorithmic outlines, examples in MATLAB, nonlinear solvers in MATLAB that can be used for production work, and chapter-ending projects.

We use the standard notation

$$F(x) = 0 \tag{1.1}$$

for systems of N equations in N unknowns. Here $F : R^N \to R^N$. We will call F the **nonlinear residual** or simply the **residual**. Rarely can the solution of a nonlinear equation be given by a closed-form expression, so iterative methods must be used to approximate the solution numerically. The output of an iterative method is a sequence of approximations to a solution.

1.1.1 Notation

In this book, following the convention in [42, 43], vectors are to be understood as column vectors. The vector x^* will denote a solution, x a potential solution, and $\{x_n\}_{n \geq 0}$ the sequence of iterates. We will refer to x_0 as the **initial iterate (not guess!)**. We will denote the ith component of a vector x by $(x)_i$ (note the parentheses) and the ith component of x_n by $(x_n)_i$. We will rarely need to refer to individual components of vectors. We will let $\partial f / \partial (x)_i$ denote the partial derivative of f with respect to $(x)_i$. As is standard [42], $e = x - x^*$ will denote the error. So, for example, $e_n = x_n - x^*$ is the error in the nth iterate.

If the components of F are differentiable at $x \in R^N$, we define the **Jacobian**

matrix $F'(x)$ by

$$F'(x)_{ij} = \frac{\partial (F)_i}{\partial (x)_j}(x).$$

Throughout the book, $\|\cdot\|$ will denote the Euclidean norm on R^N:

$$\|x\| = \left(\sum_{i=1}^{N} (x)_i^2 \right)^{1/2}.$$

1.2   Newton's Method

The methods in this book are variations of Newton's method. The Newton sequence is

$$x_{n+1} = x_n - F'(x_n)^{-1} F(x_n). \tag{1.2}$$

The interpretation of (1.2) is that we model F at the current iterate x_n with a linear function

$$M_n(x) = F(x_n) + F'(x_n)(x - x_n)$$

and let the root of M_n be the next iteration. M_n is called the **local linear model**. If $F'(x_n)$ is nonsingular, then $M_n(x_{n+1}) = 0$ is equivalent to (1.2).

Figure 1.1 illustrates the local linear model and the Newton iteration for the scalar equation

$$\arctan(x) = 0$$

with initial iterate $x_0 = 1$. We graph the local linear model

$$M_j(x) = F(x_j) + F'(x_j)(x - x_j)$$

at x_j from the point $(x_j, y_j) = (x_j, F(x_j))$ to the next iteration $(x_{j+1}, 0)$. The iteration converges rapidly and one can see the linear model becoming more and more accurate. The third iterate is visually indistinguishable from the solution. The MATLAB program `ataneg.m` creates Figure 1.1 and the other figures in this chapter for the arctan function.

The computation of a Newton iteration requires

1. evaluation of $F(x_n)$ and a test for termination,

2. approximate solution of the equation

$$F'(x_n)s = -F(x_n) \tag{1.3}$$

   for the Newton step s, and

3. construction of $x_{n+1} = x_n + \lambda s$, where the step length λ is selected to guarantee decrease in $\|F\|$.

Item 2, the computation of the Newton step, consumes most of the work, and the variations in Newton's method that we discuss in this book differ most significantly

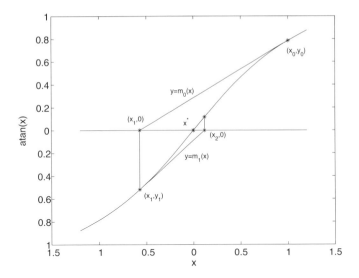

Figure 1.1. *Newton iteration for the arctan function.*

in how the Newton step is approximated. Computing the step may require evaluation and factorization of the Jacobian matrix or the solution of (1.3) by an iterative method. Not all methods for computing the Newton step require the complete Jacobian matrix, which, as we will see in Chapter 2, can be very expensive.

In the example from Figure 1.1, the step s in item 2 was satisfactory, and item 3 was not needed. The reader should be warned that attention to the step length is generally very important. One should not write one's own nonlinear solver without step-length control (see section 1.6).

1.2.1 Local Convergence Theory

The convergence theory for Newton's method [24,42,57] that is most often seen in an elementary course in numerical methods is **local**. This means that one assumes that the **initial iterate** x_0 is near a solution. The local convergence theory from [24,42,57] requires the **standard assumptions**.

Assumption 1.2.1. (standard assumptions)

1. *Equation* 1.1 *has a solution* x^*.

2. $F' : \Omega \to R^{N \times N}$ *is Lipschitz continuous near* x^*.

3. $F'(x^*)$ *is nonsingular.*

Recall that Lipschitz continuity near x^* means that there is $\gamma > 0$ (the **Lipschitz constant**) such that

$$\|F'(x) - F'(y)\| \leq \gamma \|x - y\|$$

for all x, y sufficiently near x^*.

The classic convergence theorem is as follows.

Theorem 1.1. *Let the standard assumptions hold. If x_0 is sufficiently near x^*, then the Newton sequence exists (i.e., $F'(x_n)$ is nonsingular for all $n \geq 0$) and converges to x^* and there is $K > 0$ such that*

$$\|e_{n+1}\| \leq K\|e_n\|^2 \tag{1.4}$$

for n sufficiently large.

The convergence described by (1.4), in which the error in the solution will be roughly squared with each iteration, is called **q-quadratic**. Squaring the error roughly means that the number of significant figures in the result doubles with each iteration. Of course, one cannot examine the error without knowing the solution. However, we can observe the quadratic reduction in the error computationally, if $F'(x*)$ is well conditioned (see (1.13)), because the nonlinear residual will also be roughly squared with each iteration. Therefore, we should see the exponent field of the norm of the nonlinear residual roughly double with each iteration.

In Table 1.1 we report the Newton iteration for the scalar ($N = 1$) nonlinear equation

$$F(x) = \tan(x) - x = 0, \; x_0 = 4.5. \tag{1.5}$$

The solution is $x^* \approx 4.493$.

The decrease in the function is as the theory predicts for the first three iterations, then progress slows down for iteration 4 and stops completely after that. The reason for this **stagnation** is clear: one cannot evaluate the function to higher precision than (roughly) machine unit roundoff, which in the IEEE [39, 58] floating point system is about 10^{-16}.

Table 1.1. *Residual history for Newton's method.*

n	$|F(x_n)|$
0	1.3733e–01
1	4.1319e–03
2	3.9818e–06
3	5.5955e–12
4	8.8818e–16
5	8.8818e–16

Stagnation is not affected by the accuracy in the derivative. The results reported in Table 1.1 used a forward difference approximation to the derivative with a difference increment of 10^{-6}. With this choice of difference increment, the convergence speed of the nonlinear iteration is as fast as that for Newton's method, at least for this example, until stagnation takes over. The reader should be aware that difference approximations to derivatives, while usually reliable, are often expensive and can be very inaccurate. An inaccurate Jacobian can cause many problems (see section 1.9). An analytic Jacobian can require some human effort, but can be worth it in terms of computer time and robustness when a difference Jacobian performs poorly.

One can quantify this stagnation by adding the errors in the function evaluation and derivative evaluations to Theorem 1.1. The messages of Theorem 1.2 are as follows:

- Small errors, for example, machine roundoff, in the function evaluation can lead to stagnation. This type of stagnation is usually benign and, if the Jacobian is well conditioned (see (1.13) in section 1.5), the results will be as accurate as the evaluation of F.

- Errors in the Jacobian and in the solution of the linear equation for the Newton step (1.3) will affect the speed of the nonlinear iteration, but not the limit of the sequence.

Theorem 1.2. *Let the standard assumptions hold. Let a matrix-valued function $\Delta(x)$ and a vector-valued function $\epsilon(x)$ be such that*

$$\|\Delta(x)\| < \delta_J \text{ and } \|\epsilon(x)\| < \delta_F$$

for all x near x^*. Then, if x_0 is sufficiently near x^* and δ_J and δ_F are sufficiently small, the sequence

$$x_{n+1} = x_n - (F'(x_n) + \Delta(x_n))^{-1}(F(x_n) + \epsilon(x_n))$$

is defined (i.e., $F'(x_n) + \Delta(x_n)$ is nonsingular for all n) and satisfies

$$\|e_{n+1}\| \leq \bar{K}(\|e_n\|^2 + \|\Delta(x_n)\|\|e_n\| + \|\epsilon(x_n)\|) \tag{1.6}$$

for some $\bar{K} > 0$.

We will ignore the errors in the function in the rest of this book, but one needs to be aware that stagnation of the nonlinear iteration is all but certain in finite-precision arithmetic. However, the asymptotic convergence results for exact arithmetic describe the observations well for most problems.

While Table 1.1 gives a clear picture of quadratic convergence, it's easier to appreciate a graph. Figure 1.2 is a semilog plot of **residual history**, i.e., the norm of the nonlinear residual against the iteration number. The concavity of the plot is the signature of superlinear convergence. One uses the `semilogy` command in MATLAB for this. See the file `tandemo.m`, which generated Figures 1.2 and 1.3, for an example.

1.3   Approximating the Jacobian

As we will see in the subsequent chapters, it is usually most efficient to approximate the Newton step in some way. One way to do this is to approximate $F'(x_n)$ in a

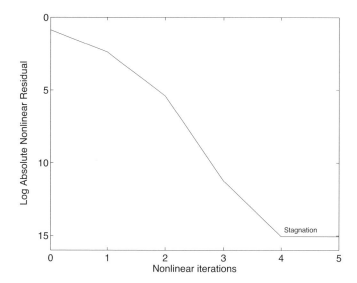

Figure 1.2. *Newton iteration for* $\tan(x) - x = 0$.

way that not only avoids computation of the derivative, but also saves linear algebra work and matrix storage.

The price for such an approximation is that the nonlinear iteration converges more slowly; i.e., more nonlinear iterations are needed to solve the problem. However, the overall cost of the solve is usually significantly less, because the computation of the Newton step is less expensive.

One way to approximate the Jacobian is to compute $F'(x_0)$ and use that as an approximation to $F'(x_n)$ throughout the iteration. This is the **chord method** or **modified Newton method**. The convergence of the chord iteration is not as fast as Newton's method. Assuming that the initial iteration is near enough to x^*, the convergence is **q-linear**. This means that there is $\rho \in (0, 1)$ such that

$$\|e_{n+1}\| \le \rho \|e_n\| \tag{1.7}$$

for n sufficiently large. We can apply Theorem 1.2 to the chord method with $\epsilon = 0$ and $\|\Delta(x_n)\| = O(\|e_0\|)$ and conclude that ρ is proportional to the initial error. The constant ρ is called the **q-factor**. The formal definition of q-linear convergence allows for faster convergence. Q-quadratic convergence is also q-linear, as you can see from the definition (1.4). In many cases of q-linear convergence, one observes that

$$\|e_{n+1}\| \approx \rho \|e_n\| \text{ or } \|F(x_{n+1})\| \approx \rho \|F(x_n)\|.$$

In these cases, q-linear convergence is usually easy to see on a semilog plot of the residual norms against the iteration number. The curve appears to be a line with slope $\approx \log(\rho)$.

The **secant method** for scalar equations approximates the derivative using a finite difference, but, rather than a forward difference, uses the most recent two iterations to form the difference quotient. So

$$x_{n+1} = x_n - \frac{F(x_n)(x_n - x_{n-1})}{F(x_n) - F(x_{n-1})}, \qquad (1.8)$$

where x_n is the current iteration and x_{n-1} is the iteration before that. The secant method must be initialized with two points. One way to do that is to let $x_{-1} = 0.99x_0$. This is what we do in our MATLAB code `secant.m`. The formula for the secant method does not extend to systems of equations ($N > 1$) because the denominator in the fraction would be a difference of vectors. We discuss one of the many generalizations of the secant method for systems of equations in Chapter 4.

The secant method's approximation to $F'(x_n)$ converges to $F'(x^*)$ as the iteration progresses. Theorem 1.2, with $\epsilon = 0$ and $\|\Delta(x_n)\| = O(\|e_{n-1}\|)$, implies that the iteration converges **q-superlinearly**. This means that either $x_n = x^*$ for some finite n or

$$\lim_{n \to \infty} \frac{\|e_{n+1}\|}{\|e_n\|} = 0. \qquad (1.9)$$

Q-superlinear convergence is hard to distinguish from q-quadratic convergence by visual inspection of the semilog plot of the residual history. The residual curve for q-superlinear convergence is concave down but drops less rapidly than the one for Newton's method.

Q-quadratic convergence is a special case of q-superlinear convergence. More generally, if $x_n \to x^*$ and, for some $p > 1$,

$$\|e_{n+1}\| = O(\|e_n\|^p),$$

we say that $x_n \to x^*$ q-superlinearly with **q-order** p.

In Figure 1.3, we compare Newton's method with the chord method and the secant method for our model problem (1.5). We see the convergence behavior that the theory predicts in the linear curve for the chord method and in the concave curves for Newton's method and the secant method. We also see the stagnation in the terminal phase.

The figure does not show the division by zero that halted the secant method computation at iteration 6. The secant method has the dangerous property that the difference between x_n and x_{n-1} could be too small for an accurate difference approximation. The division by zero that we observed is an extreme case.

The MATLAB codes for these examples are `ftst.m` for the residual; `newtsol.m`, `chordsol.m`, and `secant.m` for the solvers; and `tandemo.m` to apply the solvers and make the plots. These solvers are basic scalar codes which have user interfaces similar to those of the more advanced codes which we cover in subsequent chapters. We will discuss the design of these codes in section 1.10.

1.4 Inexact Newton Methods

Rather than approximate the Jacobian, one could instead solve the equation for the Newton step approximately. An **inexact Newton method** [22] uses as a Newton

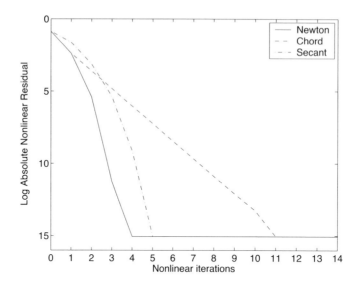

Figure 1.3. *Newton/chord/secant comparison for* $\tan(x) - x$.

step a vector s that satisfies the **inexact Newton condition**

$$\|F'(x_n)s + F(x_n)\| \leq \eta\|F(x_n)\|. \tag{1.10}$$

The parameter η (the **forcing term**) can be varied as the Newton iteration progresses. Choosing a small value of η will make the iteration more like Newton's method, therefore leading to convergence in fewer iterations. However, a small value of η may make computing a step that satisfies (1.10) very expensive. The local convergence theory [22, 42] for inexact Newton methods reflects the intuitive idea that a small value of η leads to fewer iterations. Theorem 1.3 is a typical example of such a convergence result.

Theorem 1.3. *Let the standard assumptions hold. Then there are δ and $\bar{\eta}$ such that, if $x_0 \in \mathcal{B}(\delta)$, $\{\eta_n\} \subset [0, \bar{\eta}]$, then the inexact Newton iteration*

$$x_{n+1} = x_n + s_n,$$

where

$$\|F'(x_n)s_n + F(x_n)\| \leq \eta_n\|F(x_n)\|, \tag{1.11}$$

converges q-linearly to x^*. Moreover,

- *if $\eta_n \to 0$, the convergence is q-superlinear, and*

- *if $\eta_n \leq K_\eta\|F(x_n)\|^p$ for some $K_\eta > 0$, the convergence is q-superlinear with q-order $1 + p$.*

Errors in the function evaluation will, in general, lead to stagnation of the iteration.

One can use Theorem 1.3 to analyze the chord method or the secant method. In the case of the chord method, the steps satisfy (1.11) with

$$\eta_n = O(\|e_0\|),$$

which implies q-linear convergence if $\|e_0\|$ is sufficiently small. For the secant method, $\eta_n = O(\|e_{n-1}\|)$, implying q-superlinear convergence.

Theorem 1.3 does not fully describe the performance of inexact methods in practice because the theorem ignores the method used to obtain a step that satisfies (1.10) and ignores the dependence of the cost of computing the step as a function of η.

Iterative methods for solving the equation for the Newton step would typically use (1.10) as a termination criterion. In this case, the overall nonlinear solver is called a **Newton iterative method**. Newton iterative methods are named by the particular iterative method used for the linear equation. For example, the `nsoli.m` code, which we describe in Chapter 3, is an implementation of several **Newton–Krylov** methods.

An unfortunate choice of the forcing term η can lead to very poor results. The reader is invited to try the two choices $\eta = 10^{-6}$ and $\eta = .9$ in `nsoli.m` to see this. Better choices of η include $\eta = 0.1$, the author's personal favorite, and a more complex approach (see section 3.2.3) from [29] and [42] that is the default in `nsoli.m`. Either of these usually leads to rapid convergence near the solution, but at a much lower cost for the linear solver than a very small forcing term such as $\eta = 10^{-4}$.

1.5 Termination of the Iteration

While one cannot know the error without knowing the solution, in most cases the norm of $F(x)$ can be used as a reliable indicator of the rate of decay in $\|e\|$ as the iteration progresses [42]. Based on this heuristic, we terminate the iteration in our codes when

$$\|F(x)\| \le \tau_r \|F(x_0)\| + \tau_a. \tag{1.12}$$

The relative τ_r and absolute τ_a error tolerances are both important. Using only the relative reduction in the nonlinear residual as a basis for termination (i.e., setting $\tau_a = 0$) is a poor idea because an initial iterate that is near the solution may make (1.12) impossible to satisfy with $\tau_a = 0$.

One way to quantify the utility of termination when $\|F(x)\|$ is small is to compare a relative reduction in the norm of the error with a relative reduction in the norm of the nonlinear residual. If the standard assumptions hold and x_0 and x are sufficiently near the root, then

$$\frac{\|e\|}{4\|e_0\|\kappa(F'(x^*))} \le \frac{\|F(x)\|}{\|F(x_0)\|} \le \frac{4\kappa(F'(x^*))\|e\|}{\|e_0\|}, \tag{1.13}$$

where
$$\kappa(F'(x^*)) = \|F'(x^*)\|\|F'(x^*)^{-1}\|$$
is the condition number of $F'(x^*)$ relative to the norm $\|\cdot\|$. From (1.13) we conclude that, if the Jacobian is well conditioned (i.e., $\kappa(F'(x^*))$ is not very large), then (1.12) is a useful termination criterion. This is analogous to the linear case, where a small residual implies a small error if the matrix is well conditioned.

Another approach, which is supported by theory only for superlinearly convergent methods, is to exploit the fast convergence to estimate the error in terms of the step. If the iteration is converging superlinearly, then
$$e_{n+1} = e_n + s_n = o(\|e_n\|)$$
and hence
$$s_n = -e_n + o(\|e_n\|).$$
Therefore, when the iteration is converging superlinearly, one may use $\|s_n\|$ as an estimate of $\|e_n\|$. One can estimate the current rate of convergence from above by
$$\rho_n = \|s_n\|/\|s_{n-1}\| \approx \|e_n\|/\|e_{n-1}\| \geq \|e_{n+1}\|/\|e_n\|.$$
Hence, for n sufficiently large,
$$\|e_{n+1}\| \leq \rho_n\|e_n\| \approx \|s_n\|^2/\|s_{n-1}\|.$$
So, for a superlinearly convergent method, terminating the iteration with x_{n+1} as soon as
$$\|s_n\|^2/\|s_{n-1}\| < \tau \tag{1.14}$$
will imply that $\|e_{n+1}\| < \tau$.

Termination using (1.14) is only supported by theory for superlinearly convergent methods, but is used for linearly convergent methods in some initial value problem solvers [8,61]. The trick is to estimate the q-factor ρ, say, by
$$\rho \approx \|s_n\|/\|s_{n-1}\| \text{ or } \rho \approx (\|s_n\|/\|s_0\|)^{1/n}. \tag{1.15}$$
Assuming that the estimate of ρ is reasonable, then
$$\|e_n\| - \|s_n\| \leq \|e_{n+1}\| \approx \rho\|e_n\|$$
implies that
$$\|e_{n+1}\|/\rho \approx \|e_n\| \leq \|s_n\|/(1-\rho). \tag{1.16}$$
Hence, if we terminate the iteration when
$$\|s_n\| \leq \tau(1-\rho)/\rho \tag{1.17}$$
and the estimate of ρ is an **overestimate**, then (1.16) will imply that
$$\|e_{n+1}\| \leq \rho\|s_n\|/(1-\rho) \leq \tau.$$
In practice, a safety factor is used on the left side of (1.17) to guard against an underestimate.

If, however, the estimate of ρ is much smaller than the actual q-factor, the iteration can terminate too soon. This can happen in practice if the Jacobian is ill conditioned and the initial iterate is far from the solution [45].

1.6   Global Convergence and the Armijo Rule

The requirement in the local convergence theory that the initial iterate be near the solution is more than mathematical pedantry. To see this, we apply Newton's method to find the root $x^* = 0$ of the function $F(x) = \arctan(x)$ with initial iterate $x_0 = 10$. This initial iterate is too far from the root for the local convergence theory to hold. In fact, the step

$$s = \frac{F(x_0)}{F'(x_0)} \approx \frac{1.5}{-0.01} \approx -150,$$

while in the correct direction, is far too large in magnitude.

The initial iterate and the four subsequent iterates are

$$10, -138, 2.9 \times 10^4, -1.5 \times 10^9, 9.9 \times 10^{17}.$$

As you can see, the Newton step points in the correct direction, i.e., toward $x^* = 0$, but overshoots by larger and larger amounts. The simple artifice of reducing the step by half until $\|F(x)\|$ has been reduced will usually solve this problem.

In order to clearly describe this, we will now make a distinction between the **Newton direction** $d = -F'(x)^{-1}F(x)$ and the **Newton step** when we discuss global convergence. For the methods in this book, the Newton step will be a positive scalar multiple of the Newton direction. When we talk about local convergence and are taking full steps ($\lambda = 1$ and $s = d$), we will not make this distinction and only refer to the step, as we have been doing up to now in this book.

A rigorous convergence analysis requires a bit more detail. We begin by computing the **Newton direction**

$$d = -F'(x_n)^{-1}F(x_n).$$

To keep the step from going too far, we find the smallest integer $m \geq 0$ such that

$$\|F(x_n + 2^{-m}d)\| < (1 - \alpha 2^{-m})\|F(x_n)\| \qquad (1.18)$$

and let the step be $s = 2^{-m}d$ and $x_{n+1} = x_n + 2^{-m}d$. The condition in (1.18) is called the **sufficient decrease** of $\|F\|$. The parameter $\alpha \in (0,1)$ is a small number intended to make (1.18) as easy as possible to satisfy. $\alpha = 10^{-4}$ is typical and used in our codes.

In Figure 1.4, created by `ataneg.m`, we show how this approach, called the **Armijo rule** [2], succeeds. The circled points are iterations for which $m > 1$ and the value of m is above the circle.

Methods like the Armijo rule are called **line search** methods because one searches for a decrease in $\|F\|$ along the line segment $[x_n, x_n + d]$.

The line search in our codes manages the reduction in the step size with more sophistication than simply halving an unsuccessful step. The motivation for this is that some problems respond well to one or two reductions in the step length by modest amounts (such as $1/2$) and others require many such reductions, but might do much better if a more aggressive step-length reduction (by factors of $1/10$, say)

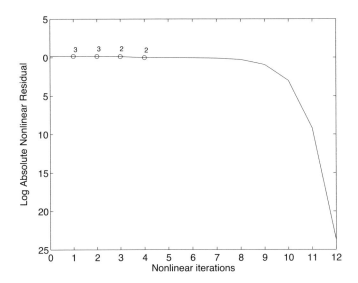

Figure 1.4. *Newton–Armijo for* arctan(x).

is used. To address this possibility, after two reductions by halving do not lead to sufficient decrease, we build a quadratic polynomial model of

$$\phi(\lambda) = \|F(x_n + \lambda d)\|^2 \tag{1.19}$$

based on interpolation of ϕ at the three most recent values of λ. The next λ is the minimizer of the quadratic model, subject to the **safeguard** that the reduction in λ be at least a factor of two and at most a factor of ten. So the algorithm generates a sequence of candidate step-length factors $\{\lambda_m\}$ with $\lambda_0 = 1$ and

$$1/10 \le \lambda_{m+1}/\lambda_m \le 1/2. \tag{1.20}$$

The norm in (1.19) is squared to make ϕ a smooth function that can be accurately modeled by a quadratic over small ranges of λ.

The line search terminates with the smallest $m \ge 0$ such that

$$\|F(x_n + \lambda_m d)\| < (1 - \alpha \lambda_m)\|F(x_n)\|. \tag{1.21}$$

In the advanced codes from the subsequent chapters, we use the three-point parabolic model from [42]. In this approach, $\lambda_1 = 1/2$. To compute λ_m for $m > 1$, a parabola is fitted to the data $\phi(0)$, $\phi(\lambda_m)$, and $\phi(\lambda_{m-1})$. λ_m is the minimum of this parabola on the interval $[\lambda_{m-1}/10, \lambda_{m-1}/2]$. We refer the reader to [42] for the details and to [24, 28, 42, 57] for a discussion of other ways to implement a line search.

1.7   A Basic Algorithm

Algorithm `nsolg` is a general formulation of an inexact Newton–Armijo iteration. The methods in Chapters 2 and 3 are special cases of `nsolg`. There is a lot of freedom in Algorithm `nsolg`. The essential input arguments are the initial iterate x, the function F, and the relative and absolute termination tolerances τ_a and τ_r. If `nsolg` terminates successfully, x will be the approximate solution on output.

Within the algorithm, the computation of the Newton direction d can be done with direct or iterative linear solvers, using either the Jacobian $F'(x)$ or an approximation of it. If you use a direct solver, then the forcing term η is determined implicitly; you do not need to provide one. For example, if you solve the equation for the Newton step with a direct method, then $\eta = 0$ in exact arithmetic. If you use an approximate Jacobian and solve with a direct method, then η is proportional to the error in the Jacobian. Knowing about η helps you understand and apply the theory, but is not necessary in practice if you use direct solvers.

If you use an iterative linear solver, then usually (1.10) is the termination criterion for that linear solver. You'll need to make a decision about the forcing term in that case (or accept the defaults from a code like `nsoli.m`, which we describe in Chapter 3). The theoretical requirements on the forcing term η are that it be safely bounded away from one (1.22).

Having computed the Newton direction, we compute a step length λ and a step $s = \lambda d$ so that the sufficient decrease condition (1.21) holds. It's standard in line search implementations to use a polynomial model like the one we described in section 1.6.

The algorithm does not cover all aspects of a useful implementation. The number of nonlinear iterations, linear iterations, and changes in the step length all should be limited. Failure of any of these loops to terminate reasonably rapidly indicates that something is wrong. We list some of the potential causes of failure in sections 1.9, 2.5, and 3.4.

Algorithm 1.1.
nsolg(x, F, τ_a, τ_r)
   Evaluate $F(x)$; $\tau \leftarrow \tau_r|F(x)| + \tau_a$.
   while $\|F(x)\| > \tau$ **do**
     Find d such that $\|F'(x)d + F(x)\| \le \eta\|F(x)\|$
     If no such d can be found, terminate with failure.
     $\lambda = 1$
     while $\|F(x + \lambda d)\| > (1 - \alpha\lambda)\|F(x)\|$ **do**
       $\lambda \leftarrow \sigma\lambda$, where $\sigma \in [1/10, 1/2]$ is computed by minimizing the polynomial model of $\|F(x_n + \lambda d)\|^2$.
     end while
     $x \leftarrow x + \lambda d$
   end while

The theory for Algorithm `nsolg` is very satisfying. If F is sufficiently smooth, η is bounded away from one (in the sense of (1.22)), the Jacobians remain well conditioned throughout the iteration, and the sequence $\{x_n\}$ remains bounded, then the iteration converges to a solution and, when near the solution, the convergence is as fast as the quality of the linear solver permits. Theorem 1.4 states this precisely, but not as generally as the results in [24, 42, 57]. The important thing that you should remember is that, for smooth F, there are only three possibilities for the iteration of Algorithm `nsolg`:

- $\{x_n\}$ will converge to a solution x^*, at which the standard assumptions hold,

- $\{x_n\}$ will be unbounded, or

- $F'(x_n)$ will become singular.

While the line search paradigm is the simplest way to find a solution if the initial iterate is far from a root, other methods are available and can sometimes overcome stagnation or, in the case of many solutions, find the solution that is appropriate to a physical problem. Trust region globalization [24, 60], pseudotransient continuation [19, 25, 36, 44], and homotopy methods [78] are three such alternatives.

Theorem 1.4. *Let $x_0 \in R^N$ and $\alpha \in (0, 1)$ be given. Assume that $\{x_n\}$ is given by Algorithm* **nsolg***, F is Lipschitz continuously differentiable,*

$$\{\eta_n\} \subset (0, \bar{\eta}] \subset (0, 1 - \alpha), \tag{1.22}$$

and $\{x_n\}$ and $\{\|F'(x_n)^{-1}\|\}$ are bounded. Then $\{x_n\}$ converges to a root x^* of F at which the standard assumptions hold, full steps ($\lambda = 1$) are taken for n sufficiently large, and the convergence behavior in the final phase of the iteration is that given by the local theory for inexact Newton methods (Theorem 1.3).

1.7.1   Warning!

The theory for convergence of the inexact Newton–Armijo iteration is only valid if $F'(x_n)$, or a very good approximation (forward difference, for example), is used to compute the step. A poor approximation to the Jacobian will cause the Newton step to be inaccurate. While this can result in slow convergence when the iterations are near the root, the outcome can be much worse when far from a solution. The reason for this is that the success of the line search is very sensitive to the direction. In particular, if x_0 is far from x^* there is **no reason** to expect the secant or chord method to converge. Sometimes methods like the secant and chord methods work fine with a line search when the initial iterate is far from a solution, but users of nonlinear solvers should be aware that the line search can fail. A good code will watch for this failure and respond by using a more accurate Jacobian or Jacobian-vector product.

Difference approximations to the Jacobian are usually sufficiently accurate. However, there are particularly hard problems [48] for which differentiation in the coordinate directions is very inaccurate, whereas differentiation in the directions of the iterations, residuals, and steps, which are natural directions for the problem, is very accurate. The inexact Newton methods, such as the Newton–Krylov methods in Chapter 3, use a forward difference approximation for Jacobian-vector products (with vectors that are natural for the problem) and, therefore, will usually (but not always) work well when far from a solution.

1.8 Things to Consider

Here is a short list of things to think about when you select and use a nonlinear solver.

1.8.1 Human Time and Public Domain Codes

When you select a nonlinear solver for your problem, you need to consider not only the computational cost (in CPU time and storage) but also **YOUR TIME**. A fast code for your problem that takes ten years to write has little value.

Unless your problem is very simple, or you're an expert in this field, your best bet is to use a public domain code. The MATLAB codes that accompany this book are a good start and can be used for small- to medium-scale production work. However, if you need support for other languages (meaning C, C++, or FORTRAN) or high-performance computing environments, there are several sources for public domain implementations of the algorithms in this book.

The Newton–Krylov solvers we discuss in Chapter 3 are at present (2003) the solvers of choice for large problems on advanced computers. Therefore, these algorithms are getting most of the attention from the people who build libraries. The SNES solver in the PETSc library [5,6] and the NITSOL [59], NKSOL [13], and KINSOL [75] codes are good implementations.

The methods from Chapter 2, which are based on direct factorizations, have received less attention recently. Some careful implementations can be found in the MINPACK and UNCMIN libraries. The MINPACK [51] library is a suite of FORTRAN codes that includes an implementation of Newton's method for dense Jacobians. The globalization is via a trust region approach [24,60] rather than the line search method we use here. The UNCMIN [65] library is based on the algorithms from [24] and includes a Newton–Armijo nonlinear equations solver. MINPACK and several other codes for solving nonlinear equations are available from the NETLIB repository at `http://www.netlib.org/`.

There is an implementation of Broyden's method in UNCMIN. This implementation is based on dense matrix methods. The MATLAB implementation that accompanies this book requires much less storage and computation.

1.8.2 The Initial Iterate

Picking an initial iterate at random (the famous "initial guess") is a bad idea. Some problems come with a good initial iterate. However, it is usually your job to create one that has as many properties of the solution as possible. Thinking about the problem and the qualitative properties of the solution while choosing the initial iterate can ensure that the solver converges more rapidly and avoids solutions that are not the ones you want.

In some applications the initial iterate is known to be good, so methods like the chord, the secant, and Broyden's method become very attractive, since the problems with the line search discussed in section 1.7.1 are not an issue. Two examples of this are implicit methods for temporal integration (see section 2.7.5), in which the initial iterate is the output of a predictor, and **nested iteration** (see section 2.8.2), where problems such as differential equations are solved on a coarse mesh and the initial iterate for the solution on finer meshes is an interpolation of the solution from a coarser mesh.

It is more common to have a little information about the solution in advance, in which case one should try to exploit those data about the solution. For example, if your problem is a discretized differential equation, make sure that any boundary conditions are reflected in your initial iterate. If you know the signs of some components of the solution, be sure that the signs of the corresponding components of the initial iterate agree with those of the solution.

1.8.3 Computing the Newton Step

If function and Jacobian evaluations are very costly, the Newton–Krylov methods from Chapter 3 and Broyden's method from Chapter 4 are worth exploring. Both methods avoid explicit computation of Jacobians, but usually require preconditioning (see sections 3.1.3, 3.2.2, and 4.3).

For very large problems, storing a Jacobian is difficult and factoring one may be impossible. Low-storage Newton–Krylov methods, such as Newton-BiCGSTAB, may be the only choice. Even if the storage is available, factorization of the Jacobian is usually a poor choice for very large problems, so it is worth considerable effort to build a good preconditioner for an iterative method. If these efforts fail and the linear iteration fails to converge, then you must either reformulate the problem or find the storage for a direct method.

A direct method is not always the best choice for a small problem, though. Integral equations, such as the example in sections 2.7.3 and 3.6.1, are one type for which iterative methods perform better than direct methods even for problems with small numbers of unknowns and dense Jacobians.

1.8.4 Choosing a Solver

The most important issues in selecting a solver are

- the size of the problem,

- the cost of evaluating F and F', and

- the way linear systems of equations will be solved.

The items in the list above are not independent.

The reader in a hurry could use the outline below and probably do well.

- If N is small and F is cheap, computing F' with forward differences and using direct solvers for linear algebra makes sense. The methods from Chapter 2 are a good choice. These methods are probably the optimal choice in terms of saving your time.

- Sparse differencing can be done in considerable generality [20, 21]. If you can exploit sparsity in the Jacobian, you will save a significant amount of work in the computation of the Jacobian and may be able to use a direct solver. The internal MATLAB code `numjac` will do sparse differencing, but requires the sparsity pattern from you. If you can obtain the sparsity pattern easily and the computational cost of a direct factorization is acceptable, a direct method is a very attractive choice.

- If N is large or computing and storing F' is very expensive, you may not be able to use a direct method.

  - If you can't compute or store F' at all, then the matrix-free methods in Chapters 3 and 4 may be your only options. If you have a good preconditioner, a Newton–Krylov code is a good start. The discussion in section 3.1 will help you choose a Krylov method.
  - If F' is sparse, you might be able to use a sparse differencing method to approximate F' and a sparse direct solver. We discuss how to do this for banded Jacobians in section 2.3 and implement a banded differencing algorithm in `nsold.m`. If you can store F', you can use that matrix to build an incomplete factorization [62] preconditioner.

1.9 What Can Go Wrong?

Even the best and most robust codes can (and do) fail in practice. In this section we give some guidance that may help you troubleshoot your own solvers or interpret hard-to-understand results from solvers written by others. These are some problems that can arise for all choices of methods. We will also repeat some of these things in subsequent chapters, when we discuss problems that are specific to a method for approximating the Newton direction.

1.9.1 Nonsmooth Functions

Most nonlinear equation codes, including the ones that accompany this book, are intended to solve problems for which F' is Lipschitz continuous. The codes will behave unpredictably if your function is not Lipschitz continuously differentiable. If, for example, the code for your function contains

- nondifferentiable functions such as the absolute value, a vector norm, or a fractional power;

- internal interpolations from tabulated data;

- control structures like *case* or *if-then-else* that govern the value returned by F; or

- calls to other codes,

you may well have a nondifferentiable problem.

If your function is close to a smooth function, the codes may do very well. On the other hand, a nonsmooth nonlinearity can cause any of the failures listed in this section.

1.9.2   Failure to Converge

The theory, as stated in Theorem 1.4, does not imply that the iteration will converge, only that nonconvergence can be identified easily. So, if the iteration fails to converge to a root, then either the iteration will become unbounded or the Jacobian will become singular.

Inaccurate function evaluation

Most nonlinear solvers, including the ones that accompany this book, assume that the errors in the evaluation are on the order of machine roundoff and, therefore, use a difference increment of $\approx 10^{-7}$ for finite difference Jacobians and Jacobian-vector products. If the error in your function evaluation is larger than that, the Newton direction can be poor enough for the iteration to fail. Thinking about the errors in your function and, if necessary, changing the difference increment in the solvers will usually solve this problem.

No solution

If your problem has no solution, then any solver will have trouble. The clear symptoms of this are divergence of the iteration or failure of the residual to converge to zero. The causes in practice are less clear; errors in programming (a.k.a. bugs) are the likely source. If F is a model of a physical problem, the model itself may be wrong. The algorithm for computing F, while technically correct, may have been realized in a way that destroys the solution. For example, internal tolerances to algorithms within the computation of F may be too loose, internal calculations based on table lookup and interpolation may be inaccurate, and if-then-else constructs can make F nondifferentiable.

If $F(x) = e^{-x}$, then the Newton iteration will diverge to $+\infty$ from any starting point. If $F(x) = x^2 + 1$, the Newton–Armijo iteration will converge to 0, the minimum of $|F(x)|$, which is not a root.

Singular Jacobian

The case where F' approaches singularity is particularly dangerous. In this case the step lengths approach zero, so if one terminates when the step is small and fails to check that F is approaching zero, one can incorrectly conclude that a root has been found. The example in section 2.7.2 illustrates how an unfortunate choice of initial iterate can lead to this behavior.

Alternatives to Newton–Armijo

If you find that a Newton–Armijo code fails for your problem, there are alternatives to line search globalization that, while complex and often more costly, can be more robust than Newton–Armijo. Among these methods are trust region methods [24, 60], homotopy [78], and pseudotransient continuation [44]. There are public domain codes for the first two of these alternatives. If these methods fail, you should see if you've made a modeling error and thus posed a problem with no solution.

1.9.3 Failure of the Line Search

If the line search reduces the step size to an unacceptably small value and the Jacobian is not becoming singular, the quality of the Newton direction is poor. We repeat the caution from section 1.7.1 that the theory for convergence of the Armijo rule depends on using the exact Jacobian. A difference approximation to a Jacobian or Jacobian-vector product is usually, but not always, sufficient.

The difference increment in a forward difference approximation to a Jacobian or a Jacobian-vector product should be a bit more than the square root of the error in the function. Our codes use $h = 10^{-7}$, which is a good choice unless the function contains components such as a table lookup or output from an instrument that would reduce the accuracy. Central difference approximations, where the optimal increment is roughly the cube root of the error in the function, can improve the performance of the solver, but for large problems the cost, twice that of a forward difference, is rarely justified. One should **scale** the finite difference increment to reflect the size of x (see section 2.3).

If you're using a direct method to compute the Newton step, an analytic Jacobian may make the line search perform much better.

Failure of the line search in a Newton–Krylov iteration may be a symptom of loss of orthogonality in the linear solver. See section 3.4.2 for more about this problem.

1.9.4 Slow Convergence

If you use Newton's method and observe slow convergence, the chances are good that the Jacobian, Jacobian-vector product, or linear solver is inaccurate. The local superlinear convergence results from Theorems 1.1 and 1.3 only hold if the correct linear system is solved to high accuracy.

If you expect to see superlinear convergence, but do not, you might try these things:

- If the errors in F are significantly larger than floating point roundoff, then increase the difference increment in a difference Jacobian to roughly the square root of the errors in the function [42].

- Check your computation of the Jacobian (by comparing it to a difference, for example).

- If you are using a sparse-matrix code to solve for the Newton step, be sure that you have specified the correct sparsity pattern.

- Make sure the tolerances for an iterative linear solver are set tightly enough to get the convergence you want. Check for errors in the preconditioner and try to investigate its quality.

- If you are using a GMRES solver, make sure that you have not lost orthogonality (see section 3.4.2).

1.9.5 Multiple Solutions

In general, there is no guarantee that an equation has a unique solution. The solvers we discuss in this book, as well as the alternatives we listed in section 1.9.2, are supported by the theory that says that either the solver will converge to a root or it will fail in some well-defined manner. No theory can say that the iteration will converge to the solution that you want. The problems we discuss in sections 2.7.3, 2.7.4, and 3.6.2 have multiple solutions.

1.9.6 Storage Problems

If your problem is large and the Jacobian is dense, you may be unable to store that Jacobian. If your Jacobian is sparse, you may not be able to store the factors that the sparse Gaussian elimination in MATLAB creates. Even if you use an iterative method, you may not be able to store the data that the method needs to converge. GMRES needs a vector for each linear iteration, for example. Many computing environments, MATLAB among them, will tell you that there is not enough storage for your job. MATLAB, for example, will print this message:

```
Out of memory. Type HELP MEMORY for your options.
```

When this happens, you can find a way to obtain more memory or a larger computer, or use a solver that requires less storage. The Newton–Krylov methods and Broyden's method are good candidates for the latter.

Other computing environments solve run-time storage problems with virtual memory. This means that data are sent to and from disk as the computation proceeds. This is called **paging** and will slow down the computation by factors of 100 or more. This is rarely acceptable. Your best option is to find a computer with more memory.

Modern computer architectures have complex memory hierarchies. The registers in the CPU are the fastest, so you do best if you can keep data in registers as long as possible. Below the registers can be several layers of cache memory. Below the cache is RAM, and below that is disk. Cache memory is faster than RAM, but much more expensive, so a cache is small. Simple things such as ordering loops to improve the locality of reference can speed up a code dramatically. You probably don't have to think about cache in MATLAB, but in FORTRAN or C, you do. The discussion of loop ordering in [23] is a good place to start learning about efficient programming for computers with memory hierarchies.

1.10   Three Codes for Scalar Equations

Three simple codes for scalar equations illustrate the fundamental ideas well. `newtsol.m`, `chordsol.m`, and `secant.m` are MATLAB implementations of Newton's method, the chord method, and the secant method, respectively, for scalar equations. They have features in common with the more elaborate codes from the rest of the book. As codes for scalar equations, they do not need to pay attention to numerical linear algebra or worry about storing the iteration history.

The Newton's method code includes a line search. The secant and chord method codes do not, taking the warning in section 1.7.1 a bit too seriously.

1.10.1 Common Features

The three codes require an initial iterate x, the function f, and relative and absolute residual tolerances *tola* and *tolr*. The output is the final result and (optionally) a history of the iteration. The history is kept in a two- or four-column *hist* array. The first column is the iteration counter and the second the absolute value of the residual after that iteration. The third and fourth, for Newton's method only, are the number of times the line search reduced the step size and the Newton sequence $\{x_n\}$. Of course, one need not keep the iteration number in the history and our codes for systems do not, but, for a simple example, doing so makes it as easy as possible to plot iteration statistics.

Each of the scalar codes has a limit of 100 nonlinear iterations. The codes can be called as follows:

```
[x, hist] = solver(x, f, tola, tolr)
```

or, if you're not interested in the history array,

```
x = solver(x, f, tola, tolr).
```

One MATLAB command will make a semilog plot of the residual history:

```
semilogy(hist(:,1),hist(:,2)).
```

1.10.2 newtsol.m

newtsol.m is the only one of the scalar codes that uses a line search. The step-length reduction is done by halving, not by the more sophisticated polynomial model based method used in the codes for systems of equations.

newtsol.m lets you choose between evaluating the derivative with a forward difference (the default) and analytically in the function evaluation. The calling sequence is

```
[x, hist] = newtsol(x, f, tola, tolr, jdiff).
```

jdiff is an optional argument. Setting $jdiff = 1$ directs newtsol.m to expect a function f with two output arguments

```
[y,yp]=f(x),
```

where $y = F(x)$ and $yp = F'(x)$. The most efficient way to write such a function is to only compute F' if it is requested. Here is an example. The function fatan.m returns the arctan function and, optionally, its derivative:

```
function [y,yp] = fatan(x)
% FATAN   Arctangent function with optional derivative
%         [Y,YP] = FATAN(X) returns Y\,=\,atan(X) and
%         (optionally) YP = 1/(1+X^2).
%
y = atan(x);
if nargout == 2
    yp = 1/(1+x^2);
end
```

The history array for `newtsol.m` has four columns. The third column is the number of times the step size was reduced in the line search. This allows you to make plots like Figure 1.4. The fourth column contains the Newton sequence.

The code below, for example, creates the plot in Figure 1.4. To use the `semilogy` to plot circles when the line search was required in this example, knowledge of the history of the iteration was needed. Here is a call to `newtsol` followed by an examination of the first five rows of the history array:

```
>> x0=10; tol=1.d-12;
>> [x,hist] = newtsol(x0, 'fatan', tol,tol);
>> hist(1:5,:)

ans =

              0     1.4711e+00              0     1.0000e+01
     1.0000e+00     1.4547e+00     3.0000e+00    -8.5730e+00
     2.0000e+00     1.3724e+00     3.0000e+00     4.9730e+00
     3.0000e+00     1.3170e+00     2.0000e+00    -3.8549e+00
     4.0000e+00     9.3921e-01     2.0000e+00     1.3670e+00
```

The third column tell us that the step size was reduced for the first through fourth iterates. After that, full steps were taken. This is the information we need to locate the circles and the numbers on the graph in Figure 1.4. Once we know that the line search is active only on iterations 1, 2, 3, and 4, we can use rows 2, 3, 4, and 5 of the history array in the plot.

```
% EXAMPLE   Draw Figure 1.4.
%
x0=10; tol=1.d-12;
[x,hist] = newtsol(x0, 'fatan', tol,tol);
semilogy(hist(:,1),abs(hist(:,2)),hist(2:5,1),abs(hist(2:5,2)),'o')
xlabel('iterations'); ylabel('function absolute values');
```

1.10.3   `chordsol.m`

`chordsol.m` approximates the Jacobian at the initial iterate with a forward difference and uses that approximation for the entire nonlinear iteration. The calling sequence is

```
[x, hist] = chordsol(x, f, tola, tolr).
```

The *hist* array has two columns, the iteration counter and the absolute value of the nonlinear residual. If you write f as you would for `newtsol.m`, with an optional second output argument, `chordsol.m` will accept it but won't exploit the analytic derivative. We invite the reader to extend `chordsol.m` to accept analytic derivatives; this is not hard to do by reusing some code from `newtsol.m`.

1.10.4 secant.m

The secant method needs two approximations to x^* to begin the iteration. `secant.m` uses the initial iterate $x_0 = x$ and then sets

$$x_{-1} = \begin{cases} 0.99x_0 & \text{if } x_0 \neq 0, \\ 0.001 & \text{if } x_0 = 0. \end{cases}$$

When stagnation takes place, a secant method code must take care to avoid division by zero in (1.8). `secant.m` does this by only updating the iteration if $x_{n-1} \neq x_n$.

The calling sequence is the same as for `chordsol.m`:

```
[x, hist] = secant(x, f, tola, tolr).
```

The three codes `newtsol.m`, `chordsol.m`, and `secant.m` were used together in `tandemo.m` to create Figure 1.3, Table 1.1, and Figure 1.2. The script begins with initialization of the solvers and calls to all three:

```
% EXAMPLE   Draw Figure 1.3.
%
%
x0=4.5; tol=1.d-20;
%
% Solve the problem three times.
%
[x,hist]=newtsol(x0,'ftan',tol,tol,1);
[x,histc]=chordsol(x0,'ftan',tol,tol);
[x,hists]=secant(x0,'ftan',tol,tol);
%
% Plot 15 iterations for all three methods.
%
maxit=15;
semilogy(hist(1:maxit,1),abs(hist(1:maxit,2)),'-',...
histc(1:maxit,1),abs(histc(1:maxit,2)),'--',...
hists(1:maxit,1),abs(hists(1:maxit,2)),'-.');
legend('Newton','Chord','Secant');
xlabel('Nonlinear iterations');
ylabel('Absolute Nonlinear Residual');
```

1.11 Projects

1.11.1 Estimating the q-order

One can examine the data in the *it_hist* array to estimate the q-order in the following way. If $x_n \to x^*$ with q-order p, then one might hope that

$$\|F(x_{n+1})\| \approx K\|F(x_n)\|^p$$

for some $K > 0$. If that happens, then, as $n \to \infty$,

$$\log(\|F(x_{n+1})\|) \approx p\log(\|F(x_n)\|)$$

and so

$$p \approx \frac{\log(\|F(x_{n+1})\|)}{\log(\|F(x_n)\|)}.$$

Hence, by looking at the *it_hist* array, we can estimate p.

This MATLAB code uses `nsold.m` to do exactly that for the functions $f(x) = x - \cos(x)$ and $f(x) = \arctan(x)$.

```
% QORDER a program to estimate the q-order
%
% Set nsold for Newton's method, tight tolerances.
%
x0 = 1.0; parms = [40,1,0]; tol = [1.d-8,1.d-8];
[x,histc] = nsold(x0,'fcos', tol,parms);
lhc=length(histc(:,2));
%
% Estimate the q-order.
%
qc = log(histc(2:lhc,1))./log(histc(1:lhc-1,1));
%
% Try it again with f(x) = atan(x).
%
[x,histt] = nsold(x0,'atan',tol,parms);
lht=length(histt(:,2));
%
% Estimate the q-order.
%
qt = log(histt(2:lht,1))./log(histt(1:lht-1,1));
```

If we examine the last few elements of the arrays *qc* and *qt* we should see a good estimate of the q-order until the iteration stagnates. The last three elements of *qc* are $3.8, 2.4$, and 2.1, as close to the quadratic convergence q-order of 2 as we're likely to see. For $f(x) = \arctan(x)$, the residual at the end is 2×10^{-24}, and the final four elements of *qt* are $3.7, 3.2, 3.2$, and 3.1. In fact, the correct q-order for this problem is 3. Why?

Apply this idea to the secant and chord methods for the example problems in this chapter. Try it for $\sin(x) = 0$ with an initial iterate of $x_0 = 3$. Are the estimated q-orders consistent with the theory? Can you explain the q-order that you observe for the secant method?

1.11.2 Singular Problems

Solve $F(x) = x^2 = 0$ with Newton's method, the chord method, and the secant method. Try the alternative iteration

$$x_{n+1} = x_n - 2F'(x_n)^{-1}F(x_n).$$

Can you explain your observations?

Chapter 2

Finding the Newton Step with Gaussian Elimination

Direct methods for solving the equation for the Newton step are a good idea if

- the Jacobian can be computed and **stored** efficiently and
- the cost of the factorization of the Jacobian is not excessive or
- iterative methods do not converge for your problem.

Even when direct methods work well, Jacobian factorization and storage of that factorization may be more expensive than a solution by iteration. However, direct methods are more robust than iterative methods and do not require your worrying about the possible convergence failure of an iterative method or preconditioning.

If the linear equation for the Newton step is solved exactly and the Jacobian is computed and factored with each nonlinear iteration (i.e., $\eta = 0$ in Algorithm `nsolg`), one should expect to see q-quadratic convergence until finite-precision effects produce stagnation (as predicted in Theorem 1.2). One can, of course, approximate the Jacobian or evaluate it only a few times during the nonlinear iteration, exchanging an increase in the number of nonlinear iterations for a dramatic reduction in the cost of the computation of the steps.

2.1   Direct Methods for Solving Linear Equations

In this chapter we solve the equation for the Newton step with Gaussian elimination. As is standard in numerical linear algebra (see [23, 32, 74, 76], for example), we distinguish between the factorization and the solve. The typical implementation of Gaussian elimination, called an **LU factorization**, factors the coefficient matrix A into a product of a permutation matrix and lower and upper triangular factors:

$$A = PLU.$$

The factorization may be simpler and less costly if the matrix has an advantageous structure (sparsity, symmetry, positivity, ...) [1, 23, 27, 32, 74, 76].

The permutation matrix reflects row interchanges that are done during the factorization to improve stability. In MATLAB, P is not explicitly referenced, but is encoded in L. For example, if

$$A = \begin{pmatrix} 4 & 6 & 6 \\ 2 & 2 & 3 \\ 7 & 8 & 10 \end{pmatrix},$$

the LU factorization

```
[l,u]=lu(A)
```

returned by the MATLAB command is

```
>> [l,u]=lu(a)
l =

   5.7143e-01    1.0000e+00             0
   2.8571e-01   -2.0000e-01    1.0000e+00
   1.0000e+00             0             0


u =

   7.0000e+00    8.0000e+00    1.0000e+01
            0    1.4286e+00    2.8571e-01
            0             0    2.0000e-01.
```

We will ignore the permutation for the remainder of this chapter, but the reader should remember that it is important. Most linear algebra software [1, 27] manages the permutation for you in some way.

The cost of an LU factorization of an $N \times N$ matrix is $N^3/3 + O(N^2)$ flops, where, following [27], we define a flop as an add, a multiply, and some address computations. The factorization is the most expensive part of the solution.

Following the factorization, one can solve the linear system $As = b$ by solving the two triangular systems $Lz = b$ and $Us = z$. The cost of the two triangular solves is $N^2 + O(N)$ flops.

2.2   The Newton–Armijo Iteration

Algorithm **newton** is an implementation of Newton's method that uses Gaussian elimination to compute the Newton step. The significant contributors to the computational cost are the computation and LU factorization of the Jacobian. The factorization can fail if, for example, F' is singular or, in MATLAB, highly ill conditioned.

Algorithm 2.1.
newton(x, F, τ_a, τ_r)

  Evaluate $F(x)$; $\tau \leftarrow \tau_r \|F(x)\| + \tau_a$.
  while $\|F(x)\| > \tau$ **do**
    Compute $F'(x)$; factor $F'(x) = LU$.
    if the factorization fails **then**
      report an error and terminate
    else
      solve $LUs = -F(x)$
    end if
    Find a step length λ using a polynomial model.
    $x \leftarrow x + \lambda s$
    Evaluate $F(x)$.
  end while

2.3 Computing a Finite Difference Jacobian

The effort in the computation of the Jacobian can be substantial. In some cases one can compute the function and the Jacobian at the same time and the Jacobian costs little more (see the example in section 2.7.3; also see section 2.5.2) than the evaluation of the function. However, if only function evaluations are available, then approximating the Jacobian by differences is the only option. As we said in Chapter 1, this usually causes no problems in the nonlinear iteration and a forward difference approximation is probably sufficient. One computes the forward difference approximation $(\nabla_h F)(x)$ to the Jacobian by columns. The jth column is

$$(\nabla_h F)(x)_j = \begin{cases} \dfrac{F(x + h\sigma_j e_j) - F(x)}{\sigma_j h}, & x_j \neq 0, \\[2ex] \dfrac{F(he_j) - F(x)}{h}, & x_j = 0. \end{cases} \tag{2.1}$$

In (2.1) e_j is the unit vector in the jth coordinate direction. The difference increment h should be no smaller than the square root of the inaccuracy in F. Each column of $\nabla_h F$ requires one new function evaluation and, therefore, a finite difference Jacobian costs N function evaluations.

    The difference increment in (2.1) should be **scaled**. Rather than simply perturb x by a difference increment h, roughly the square root of the error in F, in each coordinate direction, we multiply the perturbation to compute the jth column by

$$\max(|(x)_j|, 1) sign((x)_j),$$

with a view toward varying the correct fraction of the low-order bits in $(x)_j$. While this scaling usually makes little difference, it can be crucial if $|(x)_j|$ is very large. Note that we do not make adjustments if $|(x)_j|$ is very small because the lower limit on the size of the difference increment is determined by the error in F. For example, if evaluations of F are accurate to 16 decimal digits, the difference increment should change roughly the last 8 digits of x. Hence we use the scaled perturbation $\sigma_j h$, where

$$\sigma_j = \max(|(x)_j|, 1)sgn((x)_j). \tag{2.2}$$

In (2.2)

$$sgn(z) = \begin{cases} z/|z| & \text{if } z \neq 0, \\ 1 & \text{if } z = 0. \end{cases} \tag{2.3}$$

This is different from the MATLAB `sign` function, for which `sign(0) = 0`.

The cost estimates for a difference Jacobian change if F' is sparse, as does the cost of the factorization. If F' is sparse, one can compute several columns of the Jacobian with a single new function evaluation. The methods for doing this for general sparsity patterns [20,21] are too complex for this book, but we can illustrate the ideas with a forward difference algorithm for **banded Jacobians**.

A matrix A is **banded** with **upper bandwidth** n_u and **lower bandwidth** n_l if

$$A_{ij} = 0 \text{ if } j < i - n_l \text{ or } j > i + n_u.$$

The LU factorization of a banded matrix takes less time and less storage than that of a full matrix [23]. The cost of the factorization, when n_l and n_u are small in comparison to N, is $2Nn_ln_u(1 + o(1))$ floating point operations. The factors have at most $n_l + n_u + 1$ nonzeros. The MATLAB sparse matrix commands exploit this structure.

The Jacobian F' is banded with upper and lower bandwidths n_u and n_l if $(F)_i$ depends only on $(x)_j$ for

$$\max(1, i - n_l) \leq j \leq \min(N, i + n_u).$$

For example, if F' is tridiagonal, $n_l = n_u = 1$.

If F' is banded, then one can compute a numerical Jacobian several columns at a time. If F' is tridiagonal, then only columns 1 and 2 depend on $(x)_1$. Since $(F)_k$ for $k \geq 4$ is completely independent of any variables upon which $(F)_1$ or $(F)_2$ depend, we can differentiate F with respect to $(x)_1$ and $(x)_4$ at the same time. Continuing in this way, we can let

$$p_1 = (1, 0, 0, 1, 0, 0, 1, \ldots)^T$$

and compute

$$(D_h^1 F)(x) = \begin{cases} \dfrac{F(x + h\|x\|p_1) - F(x)}{h\|x\|}, & x \neq 0, \\[3mm] \dfrac{F(hp_1) - F(x)}{h}, & x = 0. \end{cases} \tag{2.4}$$

From $D_h^1 F$ we can recover the first, fourth, $\ldots$ columns of $\nabla_h F$ from $D_h^a F$ as follows:

$$(\nabla_h F)(x)_{i1} = (D_h^1 F)(x)_i \text{ for } 1 \le i \le 2,$$

$$(\nabla_h F)(x)_{i4} = (D_h^1 F)(x)_i \text{ for } 3 \le i \le 5,$$

$$(\nabla_h F)(x)_{i7} = (D_h^1 F)(x)_i \text{ for } 6 \le i \le 8, \tag{2.5}$$

$$\vdots$$

We can compute the remainder of the Jacobian after only two more evaluations. If we set

$$p_2 = (0, 1, 0, 0, 1, 0, 0, 1, \ldots)^T,$$

we can use formulas analogous to (2.4) and (2.5) to obtain the second, fifth, ... columns. Repeat the process with

$$p_3 = (0, 0, 1, 0, 0, 1, 0, 0, 1, \ldots)^T$$

to compute the final third of the columns. Hence a tridiagonal Jacobian can be approximated with differences using only three new function evaluations.

For a general banded matrix, the bookkeeping is a bit more complicated, but the central idea is the same. If the upper and lower bandwidths are $n_u < N$ and $n_l < N$, then $(F)_k$ depends on $(x)_1$ for $1 \le k \le 1 + n_l$. If we perturb in the first coordinate direction, we cannot perturb in any other direction that influences any $(F)_k$ that depends on $(x)_1$. Hence the next admissible coordinate for perturbation is $2 + n_l + n_u$. So we can compute the forward difference approximations of $\partial F / \partial(x)_1$ and $\partial F / \partial(x)_{2+n_u+n_u}$ with a single perturbation. Continuing in this way we define p_k for $1 \le k \le 1 + n_u + n_u$ by

$$p_k = (0, \ldots, 0, 1, 0, \ldots, 0, 1, 0, \ldots)^T \in R^N,$$

where there are $k - 1$ zeros before the first one and $n_l + n_u$ zeros between the ones. By using the vectors $\{p_k\}$ as the differencing directions, we can compute the forward difference Jacobian with $1 + n_l + n_u$ perturbations.

Our `nsold.m` solver uses this algorithm if the upper and lower bandwidths are given as input arguments. The matrix is stored in MATLAB's sparse format. When MATLAB factors a matrix in this format, it uses efficient factorization and storage methods for banded matrices.

```
function jac = bandjac(f,x,f0,nl,nu)
% BANDJAC  Compute a banded Jacobian f'(x) by forward differences.
%
% Inputs: f, x = function and point
%         f0 = f(x), precomputed function value
%         nl, nu = lower and upper bandwidth
%
n = length(x);
jac = sparse(n,n);
```

```
dv = zeros(n,1);
epsnew = 1.d-7;
%
% delr(ip)+1 = next row to include after ip in the
%              perturbation vector pt
%
% We'll need delr(1) new function evaluations.
%
% ih(ip), il(ip) = range of indices that influence f(ip)
%
for ip = 1:n
    delr(ip) = min([nl+nu+ip,n]);
    ih(ip) = min([ip+nl,n]);
    il(ip) = max([ip-nu,1]);
end
%
% Sweep through the delr(1) perturbations of f.
%
for is = 1:delr(1)
    ist = is;
%
% Build the perturbation vector.
%
    pt = zeros(n,1);
    while ist <= n
        pt(ist) = 1;
        ist = delr(ist)+1;
    end
%
% Compute the forward difference.
%
    x1 = x+epsnew*pt;
    f1 = feval(f,x1);
    dv = (f1-f0)/epsnew;
    ist = is;
%
% Fill the appropriate columns of the Jacobian.
%
    while ist <= n
    ilt = il(ist); iht = ih(ist);
    m = iht-ilt;
    jac(ilt:iht,ist) = dv(ilt:iht);
    ist = delr(ist)+1;
    end
end
```

The internal MATLAB code `numjac` is a more general finite difference Jacobian code. `numjac` was designed to work with the stiff ordinary differential equation integrators [68] in MATLAB. `numjac` will, for example, let you input a general sparsity pattern for the Jacobian and then use a sophisticated sparse differencing algorithm.

2.4   The Chord and Shamanskii Methods

If the computational cost of a forward difference Jacobian is high (F is expensive and/or N is large) and if an analytic Jacobian is not available, it is wise to amortize this cost over several nonlinear iterations. The **chord method** from section 1.3 does exactly that. Recall that the chord method differs from Newton's method in that the evaluation and factorization of the Jacobian are done only once for $F'(x_0)$. The advantages of the chord method increase as N increases, since both the N function evaluations and the $O(N^3)$ work (in the dense matrix case) in the matrix factorization are done only once. So, while the convergence is q-linear and more nonlinear iterations will be needed than for Newton's method, the overall cost of the solve will usually be much less. The chord method is the solver of choice in many codes for stiff initial value problems [3,8,61], where the Jacobian may not be updated for several time steps.

Algorithms `chord` and `shamanskii` are special cases of `nsolg`. Global convergence problems have been ignored, so the step and the direction are the same, and the computation of the step is based on an LU factorization of $F'(x)$ at an iterate that is generally not the current one.

Algorithm 2.2.
chord(x, F, τ_a, τ_r)
    Evaluate $F(x)$; $\tau \leftarrow \tau_r |F(x)| + \tau_a$.
    Compute $F'(x)$; factor $F'(x) = LU$.
    if the factorization fails **then**
        report an error and terminate
    else
        while $\|F(x)\| > \tau$ **do**
            Solve $LUs = -F(x)$.
            $x \leftarrow x + s$
            Evaluate $F(x)$.
        end while
    end if

A middle ground is the **Shamanskii method** [66]. Here the Jacobian factorization and matrix function evaluation are done after every m computations of the step.

Algorithm 2.3.
shamanskii$(x, F, \tau_a, \tau_r, m)$
  while $\|F(x)\| > \tau$ **do**
    Evaluate $F(x)$; $\tau \leftarrow \tau_r|F(x)| + \tau_a$.
    Compute $F'(x)$; factor $F'(x) = LU$.
    if the factorization fails **then**
      report an error and terminate
    end if
    for $p = 1 : m$ **do**
      Solve $LUs = -F(x)$.
      $x \leftarrow x + s$
      Evaluate $F(x)$; if $\|F(x)\| \leq \tau$ terminate.
    end for
  end while

If one counts as a complete iteration the full m steps between Jacobian computations and factorizations, the Shamanskii method converges q-superlinearly with **q-order** $m + 1$; i.e.,

$$\|x_{n+1} - x^*\| \leq K\|x_n - x^*\|^{m+1}$$

for some $K > 0$. Newton's method, of course, is the $m = 1$ case.

2.5   What Can Go Wrong?

The list in section 1.9 is complete, but it's worth thinking about a few specific problems that can arise when you compute the Newton step with a direct method. The major point to remember is that, if you use an approximation to the Jacobian, then the line search can fail. You should think of the chord and Shamanskii methods as local algorithms, to which a code will switch after a Newton–Armijo iteration has resolved any global convergence problems.

2.5.1   Poor Jacobians

The chord method and other methods that amortize factorizations over many non-linear iterations perform well because factorizations are done infrequently. This means that the Jacobians will be inaccurate, but, if the initial iterate is good, the Jacobians will be accurate enough for the overall performance to be far better than a Newton iteration. However, if your initial iterate is far from a solution, this in-accuracy can cause a **line search to fail**. Even if the initial iterate is acceptable, the convergence may be slower than you'd like. Our code `nsold.m` (see section 2.6) watches for these problems and updates the Jacobian if either the line search fails or the rate of reduction in the nonlinear residual is too slow.

2.5.2   Finite Difference Jacobian Error

The choice of finite difference increment h deserves some thought. You were warned in sections 1.9.3 and 1.9.4 that the difference increment in a forward difference approximation to a Jacobian or a Jacobian-vector product should be a bit more than the square root of the error in the function. Most codes, including ours, assume that the error in the function is on the order of floating point roundoff. If that assumption is not valid for your problem, the difference increment must be adjusted to reflect that. Check that you have scaled the difference increment to reflect the size of x, as we did in (2.1). If the components of x differ dramatically in size, consider a change of independent variables to rescale them.

Switching to centered differences can also help, but the cost of a centered difference Jacobian is very high. Another approach [49,73] uses complex arithmetic to get higher order accuracy. If F is smooth and can be evaluated for complex arguments, then you can get a second-order accurate derivative with a single function evaluation by using the formula

$$Im(F(x + ihu))/h = F'(x)u + O(h^2). \tag{2.6}$$

One should use (2.6) with some care if there are errors in F and, of course, one should scale h.

One other approach to more accurate derivatives is automatic differentiation [34]. Automatic differentiation software takes as its input a code for F and produces a code for F and F'. The derivatives are exact, but the codes are usually less efficient and larger than a hand-coded Jacobian program would be. Automatic differentiation software for C and FORTRAN is available from Argonne National Laboratory [38].

2.5.3   Pivoting

If F' is sparse, you may have the option to compute a sparse factorization without pivoting. If, for example, F' is symmetric and positive definite, this is the way to proceed. For general F', however, pivoting can be essential for a factorization to produce useful solutions. For sparse problems, the cost of pivoting can be large and it is tempting to avoid it. If line search fails and you have disabled pivoting in your sparse factorization, it's probably a good idea to re-enable it.

2.6   Using `nsold.m`

`nsold.m` is a Newton–Armijo code that uses Gaussian elimination to compute the Newton step. The calling sequence is

```
[sol, it_hist, ierr, x_hist] = nsold(x,f,tol,parms).
```

The default behavior of `nsold.m` is to try to avoid computation of the Jacobian and, if the reduction in the norm of the nonlinear residual is large enough (a factor of two), not to update the Jacobian and to reuse the factorization. This means that `nsold.m` becomes the chord method once the iteration is near the solution. The reader was warned in section 1.7.1 that this strategy could defeat the line search. `nsold.m` takes this danger into account by updating the Jacobian if the reduction in the norm of the residual is too small or if the line search fails (see section 2.6.2).

2.6.1   Input to `nsold.m`

The required input data are an initial iterate x, the function f, and the tolerances for termination. All our codes expect x and f to be column vectors of the same length.

The syntax for the function f is

```
function=f(x)
```

or

```
[function,jacobian]=f(x).
```

If it is easy for you to compute a Jacobian analytically, it is generally faster if you do that rather than let `nsold` compute the Jacobian as a full or banded matrix with a forward difference. If your Jacobian is sparse, but not banded, and you want to use the MATLAB sparse matrix functions, you **must** compute the Jacobian and store it as a MATLAB sparse matrix.

The H-equation code `heq.m` from section 2.7.3 in the software collection is a nontrivial example of a function with an optional Jacobian. The scalar function `fatan.m` from section 1.10.2 is a simpler example.

As in all our codes, the vector $tol = (\tau_a, \tau_r)$ contains the tolerances for the termination criterion (1.12).

If `nsold.m` is called with no optional arguments, then a forward difference Jacobian is computed and factored only if the ratio $\|F(x_n)\|/\|F(x_{n-1})\| > 0.5$ or the line search fails. In practice this means that the Jacobian is almost always updated in the global phase (i.e., when the iteration is far from the solution) of the iteration and that it is almost never updated in the local phase (i.e., when the iteration is near a solution that satisfies the standard assumptions).

The *parms* array controls the details of the iteration. The components of *parms* are

$$parms = [maxit, isham, rsham, jdiff, nl, nu].$$

maxit is the upper limit on the nonlinear iteration; the default is 40, which is usually enough. The Jacobian is computed and factored after every *isham* nonlinear iterations or whenever the ratio of successive norms of the nonlinear residual is larger than *rsham*. So, for example, *isham* = 1 and *rsham* = 0 is Newton's method. The default is *isham* = 1000 and *rsham* = 0.5, so the Jacobian is updated only if the decrease in the nonlinear residual is not sufficiently rapid. In this way the risk (see section 1.7.1) of using an out-of-date Jacobian when far from a solution is reduced.

The next parameter controls the computation of the Jacobian. You can leave this argument out if you want a difference Jacobian and you are not using the banded Jacobian factorization. A forward difference approximation ($jdiff = 1$) is the default. If you can provide an analytic Jacobian (using the optional second output argument to the function), set $jdiff = 0$. Analytic Jacobians almost always make the solver more efficient, but require human effort, sometimes more than is worthwhile. Automatic differentiation (see section 2.5.2) is a different way to obtain exact Jacobian information, but also requires some human and computational effort. If your Jacobian is sparse, MATLAB will automatically use a sparse factorization. If your Jacobian is banded, give the lower and upper bandwidths to `nsold.m` as the last two parameters. These can be left out for full Jacobians.

2.6.2 Output from `nsold.m`

The outputs are the solution *sol* and, optionally, a history of the iteration, an error flag, and the entire sequence $\{x_n\}$. The sequence of iterates is useful for making movies or generating figures like Figure 2.1. Be warned: asking for the iteration history, $\{x_n\}$ stored in columns of the array *x_hist*, can expend all of MATLAB's storage. One can use *x_hist* to create figures like Figure 2.1.

The error flag is useful, for example, if `nsold` is used within a larger code and one needs a test for success.

The history array *it_hist* has two columns. The first is the l^2-norm of the nonlinear residual and the second is the number of step-size reductions done in the line search.

The error flag *ierr* is 0 if the nonlinear iteration terminates successfully. The failure modes are *ierr* = 1, which means that the termination criterion is not met after *maxit* iterations, and *ierr* = 2, which means that the step length was reduced 20 times in the line search without satisfaction of the sufficient decrease condition (1.21). The limit of 20 can be changed with an internal parameter *maxarm* in the code.

2.7 Examples

The purposes of these examples are to illustrate the use of `nsold.m` and to compare the pure Newton's method with the default strategy. We provide codes for each example that call `nsold.m` twice, once with the default iteration parameters

$$parms = [40, 1000, .5, 0]$$

and once with the parameters for Newton's method

$$parms = [40, 1, 0, 0].$$

Note that the parameter $jdiff = 0$, indicating that we provide an analytic Jacobian. We invite the reader to try $jdiff = 1$. For the H-equation example in section 2.7.3 the difference Jacobian computation takes more time than the rest of the solve!

We also give simple examples of how one can use the solver from the command line.

2.7.1   Arctangent Function

This is a simple example to show how a function should be built for `nsold.m`. The function only computes a Jacobian if there are two output arguments. The line search in `nsold.m` uses the polynomial model and, therefore, the iteration history for Newton's method is a bit different from that in Figure 1.4.

With an initial iterate of $x_0 = 10$, even this small problem is difficult for the solver and the step length is reduced many times. It takes several iterations before `nsold`'s default mode stops updating the Jacobian and the two iterations begin to differ. The MATLAB code `atandemo.m` solves this problem using `nsold.m` with $\tau_a = \tau_r = 10^{-6}$ and compares the iteration histories graphically. Run the code and compare the plots yourself.

One can run the solver from the command line to get a feel for its operation and its output. In the lines below we apply Newton's method with coarse tolerances and report the solutions and iteration history. The columns in the *hist* array are the residual norms and the number of times the line search reduced the step length. The alert reader will see that the solution and the residual norm are the same to five significant figures. Why is that?

```
>> x0=10;
>> tol=[1.d-2,1.d-2];
>> params=[40, 1, 0,0];
>> [sol,hist]=nsold(x, 'fatan', tol, params);
>> sol

sol =

   9.6605e-04

>> hist

hist =

   1.4711e+00            0
   1.4547e+00    3.0000e+00
   1.3724e+00    3.0000e+00
   1.3170e+00    2.0000e+00
   9.3920e-01    2.0000e+00
   9.2507e-01            0
   8.8711e-01            0
   7.8343e-01            0
   5.1402e-01            0
   1.1278e-01            0
   9.6605e-04            0
```

2.7.2   A Simple Two-Dimensional Example

This example is from [24]. Here $N = 2$ and

$$F(x) = \begin{pmatrix} x_1^2 + x_2^2 - 2 \\ \exp(x_1 - 1) + x_2^2 - 2 \end{pmatrix}.$$

This function is simple enough for us to put the MATLAB code that computes the function and Jacobian here.

```
function [f,jac]=simple(x)
% SIMPLE simple two-dimensional problem with interesting
%        global convergence behavior
%
f=zeros(2,1);
f(1)=x(1)*x(1)+x(2)*x(2) - 2;
f(2)=exp(x(1)-1) + x(2)*x(2) - 2;
%
% Return the Jacobian if it's needed.
%
if nargout == 2
        jac=[2*x(1), 2*x(2); exp(x(1)-1), 2*x(2)];
end
```

The MATLAB code for this function is `simple.m` and the code that generated Figure 2.1 is `simpdemo.m`. In this example $\tau_a = \tau_r = 10^{-6}$. We investigated two initial iterates. For $x_0 = (2, 0.5)^T$, the step length was reduced twice on the first iteration. Full steps were taken after that. This is an interesting example because the iteration can stagnate at a point where $F'(x)$ is singular. If $x_0 = (3, 5)^T$, the line search will fail and the stagnation point will be near the $x(1)$-axis, where the Jacobian is singular.

In Figure 2.1 we plot the iteration history for both choices of initial iterate on a contour plot of $\|F\|$. The iteration that stagnates converges, but not to a root! Line search codes that terminate when the step is small should also check that the solution is an approximate root, perhaps by evaluating F (see section 1.9.2).

Here's the code that produced Figure 2.1. This is a fragment from `simpdemo.m`.

```
% SIMPDEMO
% This program solves the simple two-dimensional
% problem in Chapter 2 and makes Figure 2.1.
%
tol=[1.d-6,1.d-6];
%
% Create the mesh for the contour plot of || f ||.
%
vl=.1:.5:2; vr=2:4:40; v=[vl,vr];
v=.5:4:40;
v=[.25,.5:2:40];
```

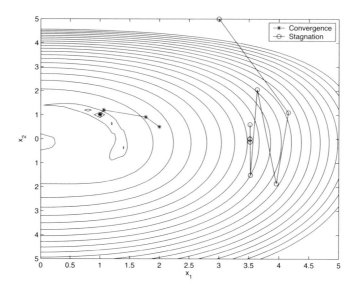

Figure 2.1. *Solution of two-dimensional example with* nsold.m.

```
xr=-5:.2:5; n=length(xr); z=zeros(n,n);
for i=1:n
for j=1:n
    w=[xr(i),xr(j)]';
    z(i,j)=norm(simple(w));
end
end
%
% Newton's method
%
params=[40, 1, 0,0];
%
% x0 is a good initial iterate.
%
x0=[2,.5]';
[sn, errsn, ierrn, x_hist]=nsold(x0, 'simple', tol, params);
%
% x1 is a poor initial iterate. The iteration from x1 will stagnate
%    at a point where F' is singular.
%
x1=[3,5]';
[sn2, errsn2, ierrn2, x_hist2]=nsold(x1, 'simple', tol, params);
%
% Draw a contour plot of || f ||.
%
```

```
figure(1)
contour(xr,xr,z,v)
hold
%
% Use the x_hist array to plot the iterations on the contour plot.
%
plot(x_hist(1,:),x_hist(2,:),'-*', x_hist2(1,:),x_hist2(2,:),'-o');
legend('Convergence','Stagnation');
xlabel('x_1');
ylabel('x_2');
axis([0 5 -5 5])
```

2.7.3   Chandrasekhar H-equation

The Chandrasekhar H-equation [15, 17] is

$$F(H)(\mu) = H(\mu) - \left(1 - \frac{c}{2}\int_0^1 \frac{\mu H(\nu)\,d\nu}{\mu + \nu}\right)^{-1} = 0. \qquad (2.7)$$

This equation arises in radiative transfer theory. There are two solutions unless $c = 0$ or $c = 1$. The algorithms and initial iterates we use in this book find the solution that is of interest physically [46]. Can you find the other one?

We will approximate the integrals by the composite midpoint rule:

$$\int_0^1 f(\mu)\,d\mu \approx \frac{1}{N}\sum_{j=1}^{N} f(\mu_j),$$

where $\mu_i = (i - 1/2)/N$ for $1 \leq i \leq N$. The resulting discrete problem is

$$F(x)_i = (x)_i - \left(1 - \frac{c}{2N}\sum_{j=1}^{N} \frac{\mu_i(x)_j}{\mu_i + \mu_j}\right)^{-1}. \qquad (2.8)$$

We will express (2.8) in a more compact form. Let A be the matrix

$$A_{ij} = \frac{c\mu_i}{2N(\mu_i + \mu_j)}.$$

Our program `heqdemo.m` for solving the H-equation stores A as a MATLAB global variable and uses it in the evaluation of both F and F'.

Once A is stored, $F(x)$ can be rapidly evaluated as

$$F(x)_i = (x)_i - (1 - (Ax)_i)^{-1}.$$

The Jacobian is given by

$$F'(x)_{ij} = \delta_{ij} - \frac{A_{ij}}{(1 - (Ax)_i)^2}.$$

Hence, once F has been computed, there is almost no new computation needed to obtain F'.

The MATLAB code for the H-equation is `heq.m`. Notice how the analytic Jacobian appears in the argument list.

```
function [h,hjac]=heq(x)
% HEQ   Chandrasekhar H-equation residual
% Jacobian uses precomputed data for fast evaluation.
%
% Be sure to store the correct data in the global array A_heq.
%
global A_heq;
n=length(x);
h=ones(n,1)-(A_heq*x);
ph=ones(n,1)./h;
h=x-ph;
if nargout==2
    hjac=(ph.*ph)*ones(1,n);
    hjac=A_heq.*hjac;
    hjac=eye(n)-hjac;
end
```

The MATLAB code `heqdemo.m` solves this equation with initial iterate $x_0 = (1,\ldots,1)^T$, $\tau_a = \tau_r = 10^{-6}$, $N = 100$, and $c = 0.9$. The output is the plot in Figure 2.2.

```
% HEQDEMO   This program creates the H-equation example in Chapter 2.
% Solve the H-equation with the default parameters in nsold and plot
% the results.
%
%
global A_heq;
c=.9;
n=100;
%
% Set the nodal points for the midpoint rule.
%
mu=1:n; mu=(mu-.5)/n; mu=mu';
%
% Form and store the kernel of the integral operator in a
% global variable.
%
cc=.5*c/n;
A_heq=ones(n,1)*mu'; A_heq=cc*A_heq'./(A_heq+A_heq');
%
tol=[1.d-6,1.d-6];
x=ones(n,1);
```

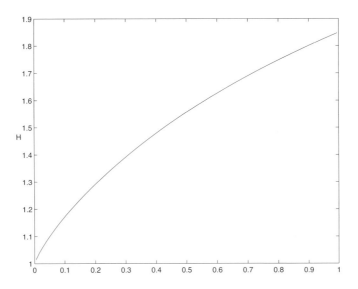

Figure 2.2. *Solution of the H-equation for* $c = 0.9$.

```
%
% Use the default parameters in nsold.m
%
[hc, errsd, ierrd]=nsold(x, 'heq', tol);
%
% Plot the results.
%
plot(gr,hc);
xlabel('\mu'); ylabel('H','Rotation',1);
```

This is a very easy problem and the Jacobian is only computed and factored once with the default settings of the parameters. Things are somewhat different with, for example, $c = 1$ and $rsham = 0.1$.

2.7.4   A Two-Point Boundary Value Problem

This example, an exercise from [3], shows how to use the banded differencing algorithm from section 2.3. We seek $v \in C^2([0, 20])$ such that

$$v''(t) + (4/t)v'(t) + (tv(t) - 1)v(t) = 0; v'(0) = v(20) = 0. \tag{2.9}$$

This problem has at least two solutions. One, $v = 0$, is not interesting, so the objective is to find a nonzero solution.

We begin by converting (2.9) to a first-order system for

$$U = \begin{pmatrix} u_1 \\ u_2 \end{pmatrix} = \begin{pmatrix} v \\ v' \end{pmatrix}.$$

The equation for U is

$$U'(t) = \left(\begin{array}{c} u_1(t) \\ u_2(t) \end{array} \right)' = G(t, U(t)) = \left(\begin{array}{c} u_2(t) \\ -(4/t)u_2(t) - (tu_1(t) - 1)u_1(t) \end{array} \right). \quad (2.10)$$

We will discretize this problem with the trapezoid rule [3, 40] on an equally spaced mesh $\{t_i\}_{i=1}^N$, where $t_i = (i-1) * h$ for $1 \leq i \leq N$ and $h = 20/(N-1)$. The discretization approximates the differential equation with the $2N - 2$ equations for $U_i \approx U(t_i)$

$$U_{i+1} - U_i = (h/2)(G(t_{i+1}, U_{i+1}) + G(t_i, U_i)) \text{ for } 1 \leq i \leq N - 1.$$

The boundary data provide the remaining two equations

$$(U_1)_2 = 0 \text{ and } (U_N)_1 = 0.$$

We can express the problem for $\{U_i\}_{i=1}^N$ as a nonlinear equation $F(x) = 0$ with a banded Jacobian of upper and lower bandwidth two by grouping the unknowns at the same point on the mesh:

$$x = (U_1^T, U_2^T, \ldots, U_N^T)^T.$$

In this way $(x)_{2i+1} \approx v(t_i)$ and $(x)_{2i} \approx v'(t_i)$. The boundary conditions are the first and last equations

$$F(x)_1 = (x)_2 = 0 \text{ and } F(x)_{2N} = (x)_{2N-1} = 0.$$

$u_1' = u_2$ is expressed in the odd components of F as

$$F(x)_{2i+1} = (x)_{2i+1} - (x)_{2i-1} - (h/2)((x)_{2i} + (x)_{2i+2})$$

for $1 \leq i \leq N - 1$.

The even components of F are the discretization of the original differential equation

$$F(x)_{2i} = (x)_{2i+2} - (x)_{2i} + (h/2)(\Phi_{i+1}(x) + \Phi_i(x)), 1 \leq i \leq N - 1.$$

Here

$$\Phi_i(x) = (4t_i^\dagger)(x)_{2i} + (t_i(x)_{2i-1} - 1)(x)_{2i-1}$$

and

$$t^\dagger = \left\{ \begin{array}{ll} 1/t & \text{if } t > 0, \\ 0 & \text{if } t = 0. \end{array} \right.$$

The MATLAB code for the nonlinear function is `bvpsys.m`. It is a direct translation of the formulas above.

```
% BVPSYS Two-point BVP for two unknown functions
%
% Problem 7.4, page 187 in [3]
```

```
%
%
function fb=bvpsys(u)
global L
n2=length(u);
fb=zeros(n2,1);
n=n2/2; h=L/(n-1);
f1=zeros(n,1); f2=zeros(n,1);
r=0:n-1; r=r'*h;
%
% Separate v and v' from their storage in u.
%
v=u(1:2:n2-1); vp=u(2:2:n2);
%
% Set the boundary conditions.
%
f1(1)=vp(1);    % v'(0) = 0
f2(n)=v(n);     % v(L) = 0;
%
f1(2:n)= v(2:n)-v(1:n-1)-h*.5*(vp(2:n)+vp(1:n-1));
%
% v'' = (4/t) v' + (t v  - 1) v
%
% The division by zero really doesn't happen. Fix it up.
%
cof=r; cof(1)=1; cof=4./cof; cof(1)=0;
%
rhs=cof.*vp + (r.*v - 1).*v;
f2(1:n-1)= vp(2:n)-vp(1:n-1) + h*.5*(rhs(2:n)+rhs(1:n-1));
fb(1:2:n2-1)=f1;
fb(2:2:n2)=f2;
```

Calling `nsold.m` is equally straightforward, but you may not get the same solution each time! Run the code `bvp2demo.m`, and then change the initial iterate to the zero vector and see what happens. `bvp2demo.m` plots v and v' as functions of t.

We can find a nonzero solution using the initial iterate

$$v(t) = e^{-t^2/10} \text{ and } v'(t) = -te^{-t^2/10}/5.$$

The solver struggles, with the line search being active for three of the nine iterations. We plot that solution in Figure 2.3. The zero solution is easy to find, too.

```
% BVP2DEMO
% This script solves the system of two-point boundary value
% problems in Chapter 2 with nsold.m.
%
```

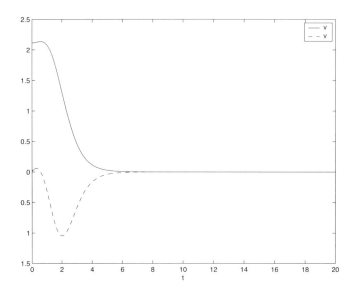

Figure 2.3. *Solution of* (2.9).

```
global L
L=20;
n=800;
u=zeros(n,1);
nh=n/2;
r=0:nh-1; h=L/(nh-1); r=r'*h;
%
% This choice of initial iterate gives the "correct" result.
% Try different initial iterates and
% watch Newton find a different solution!
%
v=exp(-r.*r*.1); vp=-.2*r.*v;
u(1:2:n-1)=v; u(2:2:n)=vp;
tol=[1.d-12,1.d-12];
%
% Use Newton's method. The upper and lower bandwidths are both 2.
%
parms=[40, 1, 0, 1, 2, 2];
[sol, it_hist, ierr] = nsold(u,'bvpsys',tol,parms);
v=sol(1:2:n-1); vp=sol(2:2:n);
it_hist
plot(r,v,'-',r,vp,'--');
xlabel('t');
legend('v','v\prime');
```

2.7.5 Stiff Initial Value Problems

Nonlinear solvers are important parts of codes for **stiff initial value problems**. In general terms [3, 67], stiffness means that either implicit methods must be used to integrate in time or, in the case of an explicit method, very small time steps must be taken.

If the problem is nonlinear, a nonlinear solver must be used at each time step. The most elementary example is the implicit Euler method. To solve the initial value problem

$$u' = G(u), u(0) = u^0 \tag{2.11}$$

with the implicit Euler method, we specify a time step δ_t and approximate the value of the solution at the mesh point $n\delta_t$ by u^n, where u^n solves the nonlinear equation

$$u^n = u^{n-1} + \delta_t G(u^n). \tag{2.12}$$

The nonlinear solver is given the function

$$F(U) = U - u^{n-1} - \delta_t G(U)$$

and an initial iterate. The initial iterate is usually either $U_0 = u^{n-1}$ or a linear predictor $U_0 = 2u^{n-1} - u^{n-2}$. In most modern codes [3, 8, 61] the termination criterion is based on small step lengths, usually something like (1.17). This eliminates the need to evaluate the function only to verify a termination condition. Similarly, the Jacobian is updated very infrequently—rarely at every time step and certainly not at every nonlinear iteration. This combination can lead to problems, but is usually very robust. The time step h depends on n in any modern initial value problem code. Hence the solver sees a different function (varying u^{n-1} and h) at each time step. We refer the reader to the literature for a complete account of how nonlinear solvers are managed in initial value problem codes and focus here on a very basic example.

As an example, consider the nonlinear parabolic problem

$$u_t = e^u + u_x x \text{ for } 0 < x < 1, 0 < t < 1 \tag{2.13}$$

with boundary data

$$u(0, t) = u(1, t) = 0 \text{ for all } 0 < t \le 1$$

and initial data

$$u(x, 0) = 0 \text{ for all } 0 \le x \le 1.$$

We solve this on a spatial mesh with width $\delta_x = 1/64$ and use a time step of $dt = 0.1$. The unknowns are approximations to $u(x_i, t_n)$ for the interior nodes $\{x_i\}_{i=1}^{63} = \{i\delta_x\}_{i=1}^{63}$ and times $\{t_i\}_{i=1}^{10} = \{i\delta_t\}_{i=1}^{10}$. Our discretization in space is the standard central difference approximation to the second derivative with homogeneous Dirichlet boundary conditions. The discretized problem is a stiff system of 63 ordinary differential equations.

For a given time step n and time increment δ_t, the components of the function F sent to `nsold.m` are given by

$$(F(U))_i = (U)_i - (u^{n-1})_i - \delta_t(e^{(U)_i} - (D_2 U)_i)$$

for $1 \le i \le N = 63$. The discrete second derivative D_2 is the tridiagonal matrix with -2 along the diagonal and 1 along the sub- and superdiagonals. All of this is encoded in the MATLAB code `ftime.m`.

```
% FTIME
% Nonlinear residual for time-dependent problem in Chapter 2
% This code has the zero boundary conditions built in.
% The time step and solution are passed as globals.
%
function ft=ftime(u)
global uold dt
%
% d2u is the numerical negative second derivative.
%
n=length(u); h=1/(n+1);
d2u=2*u;
d2u(1:n-1)=d2u(1:n-1)-u(2:n);
d2u(2:n)=d2u(2:n)-u(1:n-1);
d2u=d2u/(h^2);
%
% Nonlinear residual for implicit Euler discretization
%
ft=(u - uold) - dt * (exp(u) - d2u);
```

We pass the time step and u^{n-1} to `ftime.m` with MATLAB global variables. The code `timedep.m` integrates the initial value problem, calling `nsold.m` at each time step. The Jacobian is tridiagonal and, while computing it analytically is easy, we use the banded difference Jacobian approximation in `nsold.m`. `timedep.m` generates the time-space plot of the solution in Figure 2.4.

```
% TIMEDEP This code solves the nonlinear parabolic pde
%
% u_t = u_xx + exp(u); u(0,t) = u(1,t) = 0; u(x,0) = 0; 0 < t < 1
%
% with the backward Euler discretization. Newton's method is used
% for the nonlinear solver. The Jacobian is tridiagonal, so we
% use the banded differencing function.
%
% The value of u at the current time and the time step are passed
% to the nonlinear residual as MATLAB global variables.
%
% This problem is 1-D, so we can store the time history of the
```

```
% integration and draw a surface plot.
%
global uold dt
dt=.1;
nx=63; nt=1+1/dt;
dx=1/(nx+1);
tval=0:dt:1;
xval=0:dx:1;
%
% Use tight tolerances, Newton's method, and a tridiagonal Jacobian.
%
tol=[1.d-6,1.d-6];
parms=[40, 1, 0, 1, 1, 1];
uhist=zeros(nx+2,nt);
uold=zeros(nx,1);
for it=1:nt-1
    [unew,it_hist,ierr]=nsold(uold,'ftime',tol,parms);
    uhist(2:nx+1,it+1)=unew;
    uold=unew;
end
%
% Plot the results.
%
mesh(tval,xval,uhist)
```

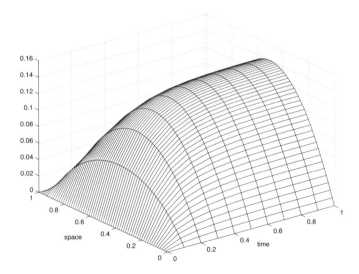

Figure 2.4. *Solution of* (2.13).

You can see from the plot that $u(x, t)$ tends to a limit as $t \to \infty$. In this case that limit is a solution of the steady-state (time-independent) equation

$$-u_{xx} = e^u \text{ for } 0 < x < 1 \tag{2.14}$$

with boundary data

$$u(0) = u(1) = 0.$$

This might give you the idea that one way to solve (2.14) would be to solve the time-dependent problem and look for convergence of u as $t \to \infty$. Of course, integrating accurately in time is a wasteful way to solve the steady-state problem, but an extension of this idea, called pseudotransient continuation, does work [19, 25, 36, 44].

2.8   Projects

2.8.1   Chandrasekhar H-equation

Solve the H-equation and plot residual histories for all of `nsold.m`, `nsoli.m`, `brsola.m`, for $c = 0.1, 0.5, 0.9, 0.99, 0.9999, 1$. Do the data in the *it_hist* array indicate super-linear convergence? Does the choice of the forcing term in `nsoli.m` affect your results?

If you suspect the convergence is q-linear, you can estimate the q-factor by examining the ratios of successive residual norms. Do this for these examples and explain your results.

If $c \neq 0, 1$, then the H-equation has two solutions [41, 52]. The one you have been computing is easy to find. Try to find the other one. This is especially entertaining for $c < 0$. For $c \notin (-\infty, 1]$, the two solutions are complex. How would you compute them?

2.8.2   Nested Iteration

Solving a differential or integral equation by nested iteration or grid sequencing means resolving the rough features of the solution of a differential or integral equation on a coarse mesh, interpolating the solution to a finer mesh, resolving on the finer mesh, and then repeating the process until you have a solution on a target mesh.

Apply this idea to some of the examples in the text, using piecewise linear interpolation to move from coarse to fine meshes. If the discretization is second-order accurate and you halve the mesh width at each level, how should you terminate the solver at each level? What kind of iteration statistics would tell you that you've done a satisfactory job?

2.9 Source Code for nsold.m

```
1    function [sol, it_hist, ierr, x_hist] = nsold(x,f,tol,parms)
2    % NSOLD  Newton-Armijo nonlinear solver
3    %
4    % Factor Jacobians with Gaussian Elimination
5    %
6    % Hybrid of Newton, Shamanskii, Chord
7    %
8    % C. T. Kelley, April 1, 2003
9    %
10   % This code comes with no guarantee or warranty of any kind.
11   %
12   % function [sol, it_hist, ierr, x_hist] = nsold(x,f,tol,parms)
13   %
14   % inputs:
15   %       initial iterate = x
16   %       function = f
17   %       tol = [atol, rtol] relative/absolute
18   %                   error tolerances
19   %       parms = [maxit, isham, rsham, jdiff, nl, nu]
20   %       maxit = maximum number of iterations
21   %               default = 40
22   %       isham, rsham: The Jacobian matrix is
23   %               computed and factored after isham
24   %               updates of x or whenever the ratio
25   %               of successive l2-norms of the
26   %               nonlinear residual exceeds rsham.
27   %
28   %       isham = -1, rsham = 0.5 is the default,
29   %       isham = 1, rsham = 0 is Newton's method,
30   %       isham = -1, rsham = 1 is the chord method,
31   %       isham = m, rsham = 1 is the Shamanskii method with
32   %               m steps per Jacobian evaluation.
33   %
34   %               The Jacobian is computed and factored
35   %               whenever the step size
36   %               is reduced in the line search.
37   %
38   %       jdiff = 1: compute Jacobians with forward differences.
39   %       jdiff = 0: a call to f will provide analytic Jacobians
40   %               using the syntax [function,jacobian] = f(x).
41   %               defaults = [40, 1000, 0.5, 1]
42   %
43   %       nl, nu: lower and upper bandwidths of a banded Jacobian.
44   %               If you include nl and nu in the parameter list,
45   %               the Jacobian will be evaluated with a banded differencing
46   %               scheme and stored as a sparse matrix.
47   %
48   %
49   %
50   % output:
51   %       sol = solution
52   %       it_hist = array of iteration history, useful for tables and plots
53   %               The two columns are the residual norm and
54   %               number of step-size reductions done in the line search.
55   %
56   %       ierr = 0 upon successful termination.
57   %       ierr = 1 if after maxit iterations
58   %               the termination criterion is not satisfied
59   %       ierr = 2 failure in the line search. The iteration
60   %               is terminated if too many step-length reductions
61   %               are taken.
62   %
63   %       x_hist = matrix of the entire iteration history.
64   %               The columns are the nonlinear iterates. This
65   %               is useful for making movies, for example, but
66   %               can consume way too much storage. This is an
67   %               OPTIONAL argument. Storage is only allocated
68   %               if x_hist is in the output argument list.
69   %
70   % internal parameter:
71   %       debug = turns on/off iteration statistics display as
72   %               the iteration progresses
73   %
74   %
75   % Here is an example. The example computes pi as a root of sin(x)
76   % with Newton's method and forward difference derivatives
77   % and plots the iteration history. Note that x_hist is not in
78   % the output list.
79   %
80   %
81   % x = 3; tol = [1.d-6, 1.d-6]; params = [40, 1, 0];
82   % [result, errs, ierr] = nsold(x, 'sin', tol, params);
83   % result
84   % semilogy(errs)
85   %
86   %
87   % Set the debug parameter, 1 turns display on, otherwise off.
88   %
89   debug = 0;
90   %
91   % Initialize it_hist and ierr and set the iteration parameters.
92   %
93   ierr = 0;
94   maxarm = 20;
95   maxit = 40;
96   isham = -1;
97   rsham = .5;
98   jdiff = 1;
99   iband = 0;
100  if nargin >= 4 & length(parms) ~= 0
```

```
101     maxit = parms(1); isham = parms(2); rsham = parms(3);
102     if length(parms) >= 4
103         jdiff = parms(4);
104     end
105     if length(parms) >= 6
106         nl = parms(5); nu = parms(6);
107         iband = 1;
108     end
109 end
110 rtol = tol(2); atol = tol(1);
111 it_hist = [];
112 n = length(x);
113 if nargout == 4, x_hist = x; end
114 fnrm = 1;
115 itc = 0;
116 %
117 % Evaluate f at the initial iterate.
118 % Compute the stop tolerance.
119 %
120 f0 = feval(f,x);
121 fnrm = norm(f0);
122 it_hist = [fnrm,0];
123 fnrmo = 1;
124 itsham = isham;
125 stop_tol = atol+rtol*fnrm;
126 %
127 % main iteration loop
128 %
129 while(fnrm > stop_tol & itc < maxit)
130 %
131 % Keep track of the ratio (rat = fnrm/frnmo)
132 % of successive residual norms and
133 % the iteration counter (itc).
134 %
135     rat = fnrm/fnrmo;
136     outstat(itc+1, :) = [itc fnrm rat];
137     fnrmo = fnrm;
138     itc = itc+1;
139 %
140 % Evaluate and factor the Jacobian
141 % on the first iteration, every isham iterate, or
142 % if the ratio of successive residual norms is too large.
143 %
144     if(itc == 1 | rat > rsham | itsham == 0 | armflag == 1)
145         itsham = isham;
146         jac_age = -1;
147         if jdiff == 1
148             if iband == 0
149                 [l, u] = diffjac(x,f,f0);
150             else
151                 jacb = bandjac(f,x,f0,nl,nu);
152                 [l,u] = lu(jacb);
153             end
154         else
155             [fv,jac] = feval(f,x);
156             [l,u] = lu(jac);
157         end
158         itsham = itsham-1;
159 %
160 % Compute the Newton direction.
161 %
162 %
163         tmp = -l\f0;
164         direction = u\tmp;
165 %
166 % Add one to the age of the Jacobian after the factors have been
167 % used in a solve. A fresh Jacobian has an age of -1 at birth.
168 %
169         jac_age = jac_age+1;
170         xold = x; fold = f0;
171         [step,iarm,x,f0,armflag] = armijo(direction,x,f0,f,maxarm);
172 %
173 % If the line search fails and the Jacobian is old, update it.
174 % If the Jacobian is fresh, you're dead.
175 %
176         if armflag == 1
177             if jac_age > 0
178                 sol = xold;
179                 x = xold; f0 = fold;
180                 disp('Armijo failure; recompute Jacobian.');
181             else
182                 disp('Complete Armijo failure.');
183                 sol = xold;
184                 ierr = 2;
185                 return
186             end
187         end
188         fnrm = norm(f0);
189         it_hist = [it_hist', [fnrm,iarm]']';
190         if nargout == 4, x_hist = [x_hist,x]; end
191         rat = fnrm/fnrmo;
192         if debug == 1, disp([itc fnrm rat]); end
193         outstat(itc+1, :) = [itc fnrm rat];
194 % end while
195 end
196 sol = x;
197 if debug == 1, disp(outstat); end
198 %
199 % On failure, set the error flag.
200 %
201 if fnrm > stop_tol, ierr = 1; end
202 %
```

```
203 %
204 %
205 function [l, u] = diffjac(x, f, f0)
206 % Compute a forward difference dense Jacobian f'(x), return lu factors.
207 %
208 % (uses dirder)
209 %
210 % C. T. Kelley, April 1, 2003
211 %
212 % This code comes with no guarantee or warranty of any kind.
213 %
214 %
215 % inputs:
216 %          x, f = point and function
217 %          f0 = f(x), preevaluated
218 %
219 n = length(x);
220 for j = 1:n
221     zz = zeros(n,1);
222     zz(j) = 1;
223     jac(:,j) = dirder(x,zz,f,f0);
224 end
225 [l, u] = lu(jac);
226 function jac = bandjac(f,x,f0,nl,nu)
227 % BANDJAC  Compute a banded Jacobian f'(x) by forward differences.
228 %
229 % Inputs: f, x = function and point
230 %         f0 = f(x), precomputed function value
231 %         nl, nu = lower and upper bandwidth
232 %
233 n = length(x);
234 jac = sparse(n,n);
235 dv = zeros(n,1);
236 epsnew = 1.d-7;
237 %
238 % delr(ip)+1 = next row to include after ip in the
239 %              perturbation vector pt
240 %
241 % We'll need delr(1) new function evaluations.
242 %
243 % ih(ip), il(ip) = range of indices that influence f(ip)
244 %
245 for ip = 1:n
246     delr(ip) = min([nl+nu+ip,n]);
247     ih(ip) = min([ip+nl,n]);
248     il(ip) = max([ip-nu,1]);
249 end
250 %
251 % Sweep through the delr(1) perturbations of f.
252 %
253 for is = 1:delr(1)
254     ist = is;
255 %
256 % Build the perturbation vector.
257 %
258     pt = zeros(n,1);
259     while ist <= n
260         pt(ist) = 1;
261         ist = delr(ist)+1;
262     end
263 %
264 % Compute the forward difference.
265 %
266     x1 = x+epsnew*pt;
267     f1 = feval(f,x1);
268     dv = (f1-f0)./epsnew;
269     ist = is;
270 %
271 % Fill the appropriate columns of the Jacobian.
272 %
273     while ist <= n
274         ilt = il(ist); iht = ih(ist);
275         m = iht-ilt;
276         jac(ilt:iht,ist) = dv(ilt:iht);
277         ist = delr(ist)+1;
278     end
279 end
280 %
281 function z = dirder(x,w,f,f0)
282 % Compute a finite difference directional derivative.
283 % Approximate f'(x) w.
284 %
285 % C. T. Kelley, April 1, 2003
286 %
287 % This code comes with no guarantee or warranty of any kind.
288 %
289 function z = dirder(x,w,f,f0)
290 %
291 % inputs:
292 %           x, w = point and direction
293 %           f = function
294 %           f0 = f(x), in nonlinear iterations
295 %                f(x) has usually been computed
296 %                before the call to dirder.
297 %
298 %
299 % Hardwired difference increment
300 epsnew = 1.d-7;
301 %
302 n = length(x);
303 %
304
```

54

```
305 % Scale the step.
306 %
307 if norm(w) == 0
308     z = zeros(n,1);
309 return
310 end
311 %
312 % Now scale the difference increment.
313 %
314 xs=(x'*w)/norm(w);
315 if xs ~= 0.d0
316     epsnew=epsnew*max(abs(xs),1.d0)*sign(xs);
317 end
318 epsnew=epsnew/norm(w);
319 %
320 % del and f1 could share the same space if storage
321 % is more important than clarity.
322 %
323 del = x+epsnew*w;
324 f1 = feval(f,del);
325 z = (f1 - f0)/epsnew;
326 %
327 % Compute the step length with the three-point parabolic model.
328 %
329 function [step,iarm,xp,fp,armflag] = armijo(direction,x,f0,f,maxarm)
330 iarm = 0;
331 sigma1 = .5;
332 alpha = 1.d-4;
333 armflag = 0;
334 xp = x; fp = f0;
335 %
336 xold = x;
337 lambda = 1; lamm = 1; lamc = lambda; iarm = 0;
338 step = lambda*direction;
339 xt = x + step;
340 ft = feval(f,xt);
341 nft = norm(ft); nf0 = norm(f0); ff0 = nf0*nf0; ffc = nft*nft; ffm = nft*nft;
342 while nft >= (1 - alpha*lambda) * nf0;
343     %
344     % Apply the three-point parabolic model.
345     %
346     if iarm == 0
347         lambda = sigma1*lambda;
348     else
349         lambda = parab3p(lamc, lamm, ff0, ffc, ffm);
350     end
351     %
352     % Update x; keep the books on lambda.
353     %
354     step = lambda*direction;
355     xt = x + step;
356     lamm = lamc;
357     lamc = lambda;
358     %
359     % Keep the books on the function norms.
360     %
361     ft = feval(f,xt);
362     nft = norm(ft);
363     ffm = ffc;
364     ffc = nft*nft;
365     iarm = iarm+1;
366     if iarm > maxarm
367         disp('Armijo failure, too many reductions');
368         armflag = 1;
369         sol = xold;
370         return;
371     end
372 end
373 xp = xt; fp = ft;
374 %
375 % end of line search
376 %
377 %
378 function lambdap = parab3p(lambdac, lambdam, ff0, ffc, ffm)
379 % Apply three-point safeguarded parabolic model for a line search.
380 %
381 % C. T. Kelley, April 1, 2003
382 %
383 %
384 % This code comes with no guarantee or warranty of any kind.
385 %
386 function lambdap = parab3p(lambdac, lambdam, ff0, ffc, ffm)
387 %
388 % input:
389 % lambdac = current step-length
390 % lambdam = previous step-length
391 % ff0 = value of || F(x_c) ||^2
392 % ffc = value of || F(x_c + lambdac d) ||^2
393 % ffm = value of || F(x_c + lambdam d) ||^2
394 %
395 % output:
396 % lambdap = new value of lambda given parabolic model
397 %
398 % internal parameters:
399 % sigma0 = 0.1, sigma1 = 0.5, safeguarding bounds for the line search
400 %
401 %
402 % Set internal parameters.
403 %
404 %
405 sigma0 = .1; sigma1 = .5;
406 %
```

```
407   % Compute coefficients of interpolation polynomial.
408   %
409   % p(lambda) = ff0 + (c1 lambda + c2 lambda^2)/d1
410   %
411   % d1 = (lambdac - lambdam)*lambdac*lambdam < 0
412   %    So if c2 > 0 we have negative curvature and default to
413   %       lambdap = sigam1 * lambda.
414   %
415   c2 = lambdam*(ffc-ff0)-lambdac*(ffm-ff0);
416   if c2 >= 0
417      lambdap = sigma1*lambdac; return
418   end
419   c1 = lambdac*lambdac*(ffm-ff0)-lambdam*lambdam*(ffc-ff0);
420   lambdap = -c1*.5/c2;
421   if lambdap < sigma0*lambdac, lambdap = sigma0*lambdac; end
422   if lambdap > sigma1*lambdac, lambdap = sigma1*lambdac; end
423
```

Chapter 3

Newton–Krylov Methods

Recall from section 1.4 that an inexact Newton method approximates the Newton direction with a vector d such that

$$\|F'(x_n)d + F(x_n)\| \leq \eta\|F(x_n)\|. \tag{3.1}$$

The parameter η is called the **forcing term**.

Newton iterative methods realize the inexact Newton condition (3.1) by applying a linear iterative method to the equation for the Newton step and terminating that iteration when (3.1) holds. We sometimes refer to this linear iteration as an **inner iteration**. Similarly, the nonlinear iteration (the while loop in Algorithm `nsolg`) is often called the **outer iteration**.

The Newton–Krylov methods, as the name suggests, use Krylov subspace-based linear solvers. The methods differ in storage requirements, cost in evaluations of F, and robustness. Our code, `nsoli.m`, includes three Krylov linear solvers: GMRES [64], BiCGSTAB [77], and TFQMR [31]. Following convention, we will refer to the nonlinear methods as Newton-GMRES, Newton-BiCGSTAB, and Newton-TFQMR.

3.1 Krylov Methods for Solving Linear Equations

Krylov iterative methods approximate the solution of a linear system $Ad = b$ with a sum of the form

$$d_k = d_0 + \sum_{j=0}^{k-1}\gamma_k A^k r_0,$$

where $r_0 = b - Ad_0$ and d_0 is the initial iterate. If the goal is to approximate a Newton step, as it is here, the most sensible initial iterate is $d_0 = 0$, because we have no a priori knowledge of the direction, but, at least in the local phase of the iteration, expect it to be small.

We express this in compact form as $d_k \in \mathcal{K}_k$, where the kth **Krylov subspace** is

$$\mathcal{K}_k = \text{span}(r_0, Ar_0, \ldots, A^{k-1}r_0).$$

Krylov methods build the iteration by evaluating matrix-vector products, in very different ways, to build an iterate in the appropriate Krylov subspace. Our `nsoli.m` code, like most implementations of Newton–Krylov methods, approximates Jacobian-vector products with forward differences (see section 3.2.1). If you find that the iteration is stagnating, you might see if an analytic Jacobian-vector product helps.

3.1.1   GMRES

The easiest Krylov method to understand is the GMRES [64] method, the default linear solver in `nsoli.m`. The kth GMRES iterate is the solution of the linear least squares problem of minimizing

$$\|b - Ad_k\|^2$$

over $\mathcal{K}_k$. We refer the reader to [23,42,64,76] for the details of the implementation, pointing out only that it is not a completely trivial task to implement GMRES well.

GMRES must accumulate the history of the linear iteration as an orthonormal basis for the Krylov subspace. This is an important property of the method because one can, and often does for large problems, exhaust the available fast memory. Any implementation of GMRES must limit the size of the Krylov subspace. GMRES(m) does this by restarting the iteration when the size of the Krylov space exceeds m vectors. `nsoli.m` has a default value of $m = 40$. The convergence theory for GMRES does not apply to GMRES(m), so the performance of GMRES(m) can be poor if m is small.

GMRES, like other Krylov methods, is often, but far from always, implemented as a **matrix-free** method. The reason for this is that only matrix-vector products, rather than details of the matrix itself, are needed to implement a Krylov method.

Convergence of GMRES

As a general rule (but not an absolute law! [53]), GMRES, like other Krylov methods, performs best if the eigenvalues of A are in a few tight clusters [16,23,42,76]. One way to understand this, keeping in mind that $d_0 = 0$, is to observe that the kth GMRES residual is in $\mathcal{K}_k$ and hence can be written as a polynomial in A applied to the residual

$$r_k = b - Ad_k = p(A)r_0 = p(A)b.$$

Here $p \in \mathcal{P}_k$, the set of kth-degree **residual polynomials**. This is the set of polynomials of degree k with $p(0) = 1$. Since the kth GMRES iteration satisfies

$$\|Ad_k - b\| \le \|Az - b\|$$

for all $z \in \mathcal{K}_k$, we must have [42]

$$\|r_k\| = \min_{p \in \mathcal{P}_k} \|p(A)r_0\|. \tag{3.2}$$

This simple fact can lead to very useful error estimates.

Here, for example, is a convergence result for diagonalizable matrices. A is **diagonalizable** if there is a nonsingular matrix V such that

$$A = V \Lambda V^{-1}.$$

Here Λ is a diagonal matrix with the eigenvalues of A on the diagonal. If A is a diagonalizable matrix and p is a polynomial, then

$$p(A) = V p(\Lambda) V^{-1}.$$

A is **normal** if the **diagonalizing transformation** V is **unitary**. In this case the columns of V are the eigenvectors of A and $V^{-1} = V^H$. Here V^H is the complex conjugate transpose of V. The reader should be aware that V and Λ can be complex even if A is real.

Theorem 3.1. *Let $A = V \Lambda V^{-1}$ be a nonsingular diagonalizable matrix. Let d_k be the kth GMRES iterate. Then, for all $\bar{p}_k \in \mathcal{P}_k$,*

$$\frac{\|r_k\|}{\|r_0\|} \leq \kappa_2(V) \max_{z \in \sigma(A)} |\bar{p}_k(z)|. \tag{3.3}$$

Proof. Let $\bar{p}_k \in \mathcal{P}_k$. We can easily estimate $\|\bar{p}_k(A)\|$ by

$$\|\bar{p}_k(A)\| \leq \|V\| \|V^{-1}\| \|\bar{p}_k(\Lambda)\| \leq \kappa_2(V) \max_{z \in \sigma(A)} |\bar{p}_k(z)|,$$

as asserted. □

Suppose, for example, that A is diagonalizable, $\kappa(V) = 100$, and all the eigenvalues of A lie in a disk of radius 0.1 centered about 1 in the complex plane. Theorem 3.1 implies (using $\bar{p}_k(z) = (1 - z)^k$) that

$$\|r_k\| \leq 100(0.1)^k = 0.1^{k-2}.$$

Hence, GMRES will reduce the residual by a factor of, say, 10^5 after seven iterations. Since reduction of the residual is the goal of the linear iteration in an inexact Newton method, this is a very useful bound. See [16] for examples of similar estimates when the eigenvalues are contained in a small number of clusters. One objective of preconditioning (see section 3.1.3) is to change A to obtain an advantageous distribution of eigenvalues.

3.1.2 Low-Storage Krylov Methods

If A is symmetric and positive definite, the conjugate gradient (CG) method [35] has better convergence and storage properties than the more generally applicable Krylov methods. In exact arithmetic the kth CG iteration minimizes

$$(x - x^*)^T A (x - x^*) = e^T A e = \|e\|_A^2$$

over the kth Krylov subspace. The symmetry and positivity can be exploited so that the storage requirements do not grow with the number of iterations.

A tempting idea is to multiply a general system $Ax = b$ by A^T to obtain the **normal equations** $A^T A x = A^T b$ and then apply CG to the new problem, which has a symmetric positive definite coefficient matrix $A^T A$. This approach, called CGNR, has the disadvantage that the condition number of $A^T A$ is the square of that of A, and hence the convergence of the CG iteration can be far too slow. A similar approach, called CGNE, solves $A A^T z = b$ with CG and then sets $x = A^T z$. Because the condition number is squared and a transpose-vector multiplication is needed, CGNR and CGNE are used far less frequently than the other low-storage methods.

The need for a transpose-vector multiplication is a major problem unless one wants to store the Jacobian matrix. It is simple (see section 3.2.1) to approximate a Jacobian-vector product with a forward difference, but no matrix-free way to obtain a transpose-vector product is known.

Low-storage alternatives to GMRES that do not need a transpose-vector product are available [31, 42, 77] but do not have the robust theoretical properties of GMRES or CG. Aside from GMRES(m), two such low-storage solvers, BiCGSTAB [77] and TFQMR [31], can be used in `nsoli.m`.

We refer the reader to [31, 33, 42, 77] for detailed descriptions of these methods. If you consider BiCGSTAB and TFQMR as solvers, you should be aware that, while both have the advantage of a fixed storage requirement throughout the linear iteration, there are some problems.

Either method can **break down**; this means that the iteration will cause a division by zero. This is not an artifact of the floating point number system but is intrinsic to the methods. While GMRES(m) can also fail to converge, that failure will manifest itself as a stagnation in the iteration, not a division by zero or an overflow.

The number of linear iterations that BiCGSTAB and TFQMR need for convergence can be roughly the same as for GMRES, but each linear iteration needs two matrix-vector products (i.e., two new evaluations of F).

GMRES(m) should be your first choice. If, however, you cannot allocate the storage that GMRES(m) needs to perform well, one of BiCGSTAB or TFQMR may solve your problem. If you can store the Jacobian, or can compute a transpose-vector product in an efficient way, and the Jacobian is well conditioned, applying CG iteration to the normal equations can be a good idea. While the cost of a single iteration is two matrix-vector products, convergence, at least in exact arithmetic, is guaranteed [33, 42].

3.1.3   Preconditioning

Preconditioning the matrix A means multiplying A from the right, left, or both sides by a **preconditioner** M. One does this with the expectation that systems with the coefficient matrix MA or AM are easier to solve than those with A. Of course, preconditioning can be done in a matrix-free manner. One needs only a function that performs a preconditioner-vector product.

Left preconditioning multiplies the equation $As = b$ on both sides by M to obtain the **preconditioned system** $MAx = Mb$. One then applies the Krylov method to the preconditioned system. If the condition number of MA is really smaller than that of A, the residual of the preconditioned system will be a better reflection of the error than that of the original system. One would hope so, since the preconditioned residual will be used to terminate the linear iteration.

Right preconditioning solves the system $AMy = b$ with the Krylov method. Then the solution of the original problem is recovered by setting $x = My$. Right preconditioning has the feature that the residual upon which termination is based is the residual for the original problem.

Two-sided preconditioning replaces A with $M_{left}AM_{right}$.

A different approach, which is integrated into some initial value problem codes [10, 12], is to pretend that the Jacobian is banded, even if it isn't, and to use Jacobian-vector products and the forward difference method for banded Jacobians from section 2.3 to form a banded approximation to the Jacobian. One factors the banded approximation and uses that factorization as the preconditioner.

3.2 Computing an Approximate Newton Step

3.2.1 Jacobian-Vector Products

For nonlinear equations, the Jacobian-vector product is easy to approximate with a forward difference directional derivative. The forward difference directional derivative at x in the direction w is

$$D_h F(x:w) = \begin{cases} 0, & w = 0, \\ \|w\| \dfrac{F(x + \sigma(x,w)hw/\|w\|) - F(x)}{\sigma(x,w)h}, & w \neq 0. \end{cases} \tag{3.4}$$

The scaling is important. We first scale w to be a unit vector and take a numerical directional derivative in the direction $w/\|w\|$. If h is roughly the square root of the error in F, we use a difference increment in the forward difference to make sure that the appropriate low-order bits of x are perturbed. So we multiply h by

$$\sigma(x,w) = \max(|x^T w|, \|w\|)sgn(x^T w)/\|w\|.$$

The same scaling was used in the forward difference Jacobian in (2.1). Remember not to use the MATLAB `sign` function for sgn, which is defined by (2.3).

3.2.2 Preconditioning Nonlinear Equations

Our code `nsoli.m` expects you to incorporate preconditioning into F. The reason for this is that the data structures and algorithms for the construction and application of preconditioners are too diverse to all fit into a nonlinear solver code.

To precondition the equation for the Newton step from the left, one simply applies `nsoli.m` to the preconditioned nonlinear problem

$$G(x) = MF(x) = 0.$$

The equation for the Newton step for G is

$$G'(x)s = MF'(x)s = -G(x) = -MF(x),$$

which is the left-preconditioned equation for the Newton step for F.

If we set $x = My$ and solve

$$G(y) = F(My) = 0$$

with Newton's method, then the equation for the step is

$$G'(y) = F'(My)M\tilde{s} = -G(y) = -F(My),$$

which is the right-preconditioned equation for the step. To recover the step s in x we might use $s = M\tilde{s}$ or, equivalently, $x_+ = M(y_+ + \tilde{s})$, but it's simpler to solve $G(y) = 0$ to the desired accuracy and set $x = My$ at the end of the nonlinear solve. As in the linear case, the nonlinear residual is the same as that for the original problem.

Left or Right Preconditioning?

There is no general rule for choosing between left and right preconditioning. You should keep in mind that the two approaches terminate the nonlinear iteration differently, so you need to decide what you're interested in. Linear equations present us with exactly the same issues.

   Left preconditioning will terminate the iteration when $\|MF(x)\|$ is small. If M is a good approximation to $F'(x^*)^{-1}$, then

$$MF(x) \approx MF'(x^*)(x - x^*) \approx x - x^*$$

and this termination criterion captures the actual error. On the other hand, **right preconditioning**, by terminating when $\|F(x)\|$ is small, captures the behavior of the residual, responding to the problem statement "Make $\|F\|$ small," which is often the real objective.

3.2.3   Choosing the Forcing Term

The approach in [29] changes the forcing term η in (3.1) as the nonlinear iteration progresses. The formula is complex and motivated by a lengthy story, which we condense from [42]. The overall goal in [29] is to solve the linear equation for the Newton step to just enough precision to make good progress when far from a solution, but also to obtain quadratic convergence when near a solution. One might base a choice of η on residual norms; one way to do this is

$$\eta_n^{Res} = \gamma \|F(x_n)\|^2 / \|F(x_{n-1})\|^2,$$

where $\gamma \in (0, 1]$ is a parameter. If η_n^{Res} is bounded away from 1 for the entire iteration, the choice $\eta_n = \eta_n^{Res}$ will do the job, assuming we make a good choice for η_0. To make sure that η_n stays well away from one, we can simply limit its maximum size. Of course, if η_n is too small in the early stage of the iteration, then the linear equation for the Newton step can be solved to far more precision than is really needed. To protect against **oversolving**, a method of **safeguarding** was proposed in [29] to avoid volatile decreases in η_n. The idea is that if η_{n-1} is sufficiently large, we do not let η_n decrease by too much; [29] suggests limiting the decrease to a factor of η_{n-1}.

After taking all this into account, one finally arrives at [42]

$$\eta_n = \min(\eta_{max}, \max(\eta_n^{Safe}, 0.5\tau_t/\|F(x_n)\|)). \tag{3.5}$$

The term

$$\tau_t = \tau_a + \tau_r \|F(x_0)\|$$

is the termination tolerance for the nonlinear iteration and is included in the formula to prevent oversolving on the final iteration. η_{max} is an upper limit on the forcing term and

$$\eta_n^{Safe} = \begin{cases} \eta_{max}, & n = 0, \\[2mm] \min(\eta_{max}, \eta_n^{Res}), & n > 0, \gamma\eta_{n-1}^2 \leq 0.1, \\[2mm] \min(\eta_{max}, \max(\eta_n^{Res}, \gamma\eta_{n-1}^2)), & n > 0, \gamma\eta_{n-1}^2 > 0.1, \end{cases} \tag{3.6}$$

In [29] the choices $\gamma = 0.9$ and $\eta_{max} = 0.9999$ are used. The defaults in `nsoli.m` are $\gamma = 0.9$ and $\eta_{max} = 0.9$.

3.3   Preconditioners

This section is not an exhaustive account of preconditioning and is only intended to point the reader to the literature.

Ideally the preconditioner should be close to the inverse of the Jacobian. In practice, one can get away with far less. If your problem is a discretization of an elliptic differential equation, then the inverse of the high-order part of the differential operator (with the correct boundary conditions) is an excellent preconditioner [50]. If the high-order term is linear, one might be able to compute the preconditioner-vector product rapidly with, for example, a fast transform method (see section 3.6.3) or a multigrid iteration [9]. Multigrid methods exploit the smoothing properties of the classical stationary iterative methods by mapping the equation through a sequence of grids. When multigrid methods are used as a solver, it can often be shown that a solution can be obtained at a cost of $O(N)$ operations, where N is the number of unknowns. Multigrid implementation is difficult and a more typical application is to use a single multigrid iteration (for the high-order term) as a preconditioner.

Domain decomposition preconditioners [72] approximate the inverse of the high-order term (or the entire operator) by subdividing the geometric domain of the differential operator, computing the inverses on the subdomains, and combining those inverses. When implemented in an optimal way, the condition number of the preconditioned matrix is independent of the discretization mesh size.

Algebraic preconditioners use the sparsity structure of the Jacobian matrix. This is important, for example, for problems that do not come from discretizations of differential or integral equations or for discretizations of differential equations on unstructured grids, which may be generated by computer programs.

An example of such a preconditioner is **algebraic multigrid**, which is designed for discretized differential equations on unstructured grids. Algebraic multigrid attempts to recover geometric information from the sparsity pattern of the Jacobian and thereby simulate the intergrid transfers and smoothing used in a conventional geometric multigrid preconditioner.

Another algebraic approach is **incomplete factorization** [62,63]. Incomplete factorization preconditioners compute a factorization of a sparse matrix but do not store those elements in the factors that are too small or lie outside a prescribed sparsity pattern. These preconditioners require that the Jacobian be stored as a sparse matrix. The MATLAB commands `luinc` and `cholinc` implement incomplete LU and Cholesky factorizations.

3.4   What Can Go Wrong?

Any problem from section 1.9, of course, can arise if you solve linear systems by iteration. There are a few problems that are unique to Newton iterative methods. The symptoms of these problems are unexpectedly slow convergence or even failure/stagnation of the nonlinear iteration.

3.4.1   Failure of the Inner Iteration

When the linear iteration does not satisfy the inexact Newton condition (3.1) and the limit on linear iterations has been reached, a sensible response is to warn the user and return the step to the nonlinear iteration. Most codes, including `nsoli.m`, do this. While it is likely that the nonlinear iteration will continue to make progress, convergence is not certain and one may have to allow the linear solver more iterations, use a different linear solver, or, in extreme cases, find enough storage to use a direct solver.

3.4.2   Loss of Orthogonality

GMRES and CG exploit orthogonality of the Krylov basis to estimate the residual and, in the case of CG, conserve storage. In finite-precision arithmetic this orthogonality can be lost and the estimate of the residual in the iteration can be poor. The iteration could terminate prematurely because the estimated residual satisfies (3.1) while the true residual does not. This is a much more subtle problem than failure to converge because the linear solver can report success but return an inaccurate and useless step.

The GMRES code in `nsoli.m`, like the ones based on the GMRES solver in [11], tests for loss of orthogonality and tries to correct it. We refer the reader to [42] for the details. You have the option in most GMRES codes of forcing the iteration to maintain the orthogonality of the Krylov basis at a cost of doubling the number of scalar products in the linear iteration.

3.5   Using nsoli.m

nsoli.m is a Newton–Krylov code that uses one of several Krylov methods to satisfy the inexact Newton condition (3.1). nsoli.m expects the preconditioner to be part of the nonlinear function, as described in section 3.1.3.

The calling sequence is similar to that for nsold.m:

[sol, it_hist, ierr, x_hist] = nsoli(x, f, tol, parms).

3.5.1   Input to nsoli.m

The required data for nsoli.m are x, the function f, and the tolerances for termination. The vector $tol = (\tau_a, \tau_r)$ contains the tolerances for the termination criterion (1.12). These are the same as for nsold.m (see section 2.6.1).

x and f must be column vectors of the same length. The syntax for f is

function = f(x).

The *parms* array is more complex than that for nsold.m. The components are
$$parms = [maxit, maxitl, etamax, lmeth, restart\_limit].$$

maxit is the upper limit on the nonlinear iterations, as it is in all our codes. The default is 40. *maxitl* is the maximum number of linear iterations per nonlinear iteration, except for in GMRES(m), where it is the maximum number of iterations before a restart. The default is 40.

etamax controls the linear tolerance in the inexact Newton condition (3.1). This parameter has a dual role. If *etamax* < 0, then $\eta = |etamax|$. If *etamax* > 0, then η is determined by (3.5). The default is *etamax* $= 0.9$.

The choice of Krylov method is governed by the parameter *lmeth*. GMRES (*lmeth* = 1) is the default. The other alternatives are GMRES(m) (*lmeth* = 2), BiCGSTAB (*lmeth* = 3), and TFQMR (*lmeth* = 4). The values of *maxit*, *maxitl*, and η must be set if you change the value of *lmeth*. If GMRES(m) is the linear solver, one must also specify the total number of restarts in *restart_limit*. The default is 20, which means that GMRES(m) is allowed $20 \times 40 = 800$ linear iterations per nonlinear iteration.

3.5.2   Output from nsoli.m

Like nsold.m, the outputs are the solution *sol* and, optionally, a history of the iteration, an error flag, and the entire sequence $\{x_n\}$. The sequence of iterates is useful for making movies or generating figures like Figure 2.1. Don't ask for the sequence $\{x_n\}$ unless you have enough storage for this array. For large problems, asking for the iteration history $\{x_n\}$ by including x_hist in the argument list can expend all of MATLAB's storage. The code `ozmovie.m` in the directory for this chapter is an example of how to use the sequence of iterates to make a movie.

The history array it_hist has three columns. The first is the Euclidean norm of the nonlinear residual $\|F(x)\|$, the second is the cumulative number of calls to F, and the third is the number of step-size reductions done in the line search.

The error flag $ierr$ is 0 if the nonlinear iteration terminates successfully. The failure modes are $ierr = 1$, which means that the termination criterion is not met after $maxit$ iterations, and $ierr = 2$, which means that the step length was reduced 20 times in the line search without satisfaction of the sufficient decrease condition (1.21). The limit of 20 can be changed with an internal parameter $maxarm$ in the code.

3.6 Examples

Often iterative methods are faster than direct methods even if the Jacobian is small and dense. That's the case with the H-equation in our first example in section 3.6.1. If the Jacobian is too expensive to compute and store, as is the case with the other two examples, factoring the Jacobian is not an option.

3.6.1 Chandrasekhar H-equation

To get started, we solve the H-equation (2.7) on a mesh of 100 points with a variety of Newton–Krylov methods and compare the performance by plotting the relative nonlinear residual $\|F(x_n)\|/\|F(x_0)\|$ against the number of calls to F. The initial iterate was the vector `ones(100,1)` and $\tau_a = \tau_r = 10^{-8}$.

The code `heqkdemo.m` calls `nsoli.m` with three sets of the parameter array

$$parms = [40, 40, 0.9, lmeth]$$

with $lmeth = 1, 3, 4$ for Newton-GMRES, Newton-BiCGSTAB, and Newton-TFQMR, respectively. Note that the values of $maxit$, $maxitl$, and η are the defaults but must be included if $lmeth$ is to be varied. The forcing term is computed using (3.5).

`heqkdemo.m` draws two figures, one that plots the residual against the nonlinear iteration count and another, shown in Figure 3.1, with the number of calls to F on the horizontal axis. In this way we can better estimate the total cost and see, for this example, that GMRES requires fewer calls to F than the other two linear solvers and therefore is preferable if the storage that GMRES needs is available. TFQMR and BiCGSTAB need two Jacobian-vector products for each linear iteration, which accounts for their added cost.

Generating such a plot is simple. This MATLAB fragment does the job with Newton-TFQMR and an initial iterate of $H = 1$ to produce a plot of the norm of the nonlinear residual against the number of function evaluations (the dot-dash curve in Figure 3.1).

```
% NEWTON-TFQMR SOLUTION OF H-EQUATION
% Call nsoli to solve the H-equation with Newton-TFQMR.
```

```
%
x=ones(100,1);
tol=[1.d-8,1.d-8];
parms = [40,40,.9,4];
[sol, it_hist, ierr] = nsoli(x,'heq',tol,parms);
%
% Plot a residual history.
%
semilogy(it_hist(:,2),it_hist(:,1)/it_hist(1,1));
```

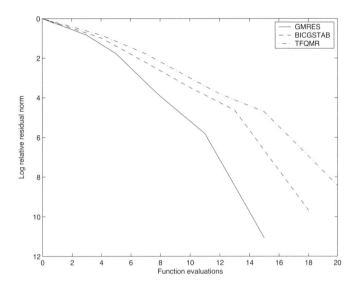

Figure 3.1. *Nonlinear residual versus calls to F.*

3.6.2   The Ornstein–Zernike Equations

This example is taken from [7, 18, 56]. The problem is an integral equation coupled with an algebraic constraint. After approximating the integral with the trapezoid rule, we find that the function can be most efficiently evaluated with a fast Fourier transform (FFT), making the computation of the Newton step with a direct method impractical. The unknowns are two continuous functions h and c defined on $0 \leq r < \infty$. It is standard to truncate the computational domain and seek $h, c \in C[0, L]$. For this example $L = 9$.

In their simplest isotropic form the Ornstein–Zernike (OZ) equations are a system consisting of an integral equation

$$F(h,c)(r) = h(r) - c(r) - \rho(h * c)(r) = 0, \qquad (3.7)$$

where

$$(h * c)(r) = \int c(\|\mathbf{r} - \mathbf{r}'\|)h(\|\mathbf{r}'\|)d\mathbf{r}' \qquad (3.8)$$

and the integral is over R^3. Here ρ is a parameter.

The nonlinear algebraic constraint is

$$G(h,c)(r) = \exp(-\beta u(r) + h(r) - c(r)) - h(r) - 1 = 0 \text{ for all } 0 \le r \le R. \quad (3.9)$$

In (3.9)

$$u(r) = 4\epsilon \left(\left(\frac{\sigma}{r}\right)^{12} - \left(\frac{\sigma}{r}\right)^6 \right) \quad (3.10)$$

and β, ϵ, and σ are parameters. For this example we use

$$\beta = 10, \rho = 0.2, \epsilon = 0.1, \text{ and } \sigma = 2.0.$$

The convolution $h * c$ in (3.7) can be computed with only one-dimensional integrals using the spherical-Bessel transform. If h decays sufficiently rapidly, as we assume, we define

$$\hat{h}(k) = \mathcal{H}(h)(k) = 4\pi \int_0^\infty \frac{\sin(kr)}{kr} h(r) r^2 dr$$

and

$$h(r) = \mathcal{H}^{-1}(\hat{h})(r) = \frac{1}{2\pi^2} \int_0^\infty \frac{\sin(kr)}{kr} \hat{h}(k) k^2 dk.$$

We compute $h * c$ by discretizing the formula

$$h * c = \mathcal{H}^{-1}(\hat{h}\hat{c}), \quad (3.11)$$

where $\hat{h}\hat{c}$ is the pointwise product of functions.

Discrete Problem

We will approximate the values of h and c on the mesh

$$\Omega_\delta = \{r_i^\delta\}_{i=1}^N,$$

where $\delta = L/(N-1)$ is the mesh width and $r_i^\delta = (i-1)\delta$.

To approximate $h * c$, we begin by discretizing frequency in a way that allows us to use the FFT to evaluate the convolution. Let $k_j = (j-1)\delta_k$, where $\delta_k = \pi\delta/(N-1)$. We define, for $2 \le j \le N-1$,

$$\hat{v}_j = \mathcal{H}(v)(k_j)$$

$$= \frac{4\pi\delta^2}{(j-1)\delta_k} \sum_{i=2}^{N-1} (i-1)v_i \sin((i-1)(j-1)\delta_k\delta) \quad (3.12)$$

$$= \frac{4\pi\delta^3(N-1)}{j-1} \sum_{i=2}^{N-1} (i-1)v_i \sin((i-1)(j-1)\pi/(N-1)).$$

Then, for $2 \le i \le N-1$,

$$\mathcal{H}^{-1}(\hat{v})_i = \frac{1}{2(i-1)\pi\delta^3} \sum_{j=2}^{N-1} (j-1)\hat{v}_j \sin((i-1)(j-1)\pi/(N-1)). \quad (3.13)$$

Finally, we define, for $2 \leq i \leq N - 1$,

$$(u * v)_i = \mathcal{H}^{-1}(\hat{u}\hat{v})_i,$$

where $\hat{u}\hat{v}$ denotes the componentwise product. We set $(u * v)_N = 0$ and define $(u * v)_1$ by linear interpolation as

$$(u * v)_1 = 2(u * v)_2 - (u * v)_3.$$

The sums in (3.12) and (3.13) can be done with a **fast sine transform** using the **MATLAB FFT**. To compute the sums

$$l_i = \sum_{i=1}^{N-1} \sin(ij\pi/N) f_j \tag{3.14}$$

for $1 \leq i \leq N - 1$ one can use the MATLAB code `lsint` to compute `l = lsint(f)`. The sine transform code is as follows:

```
% LSINT
% Fast sine transform with MATLAB's FFT
%
function lf=lsint(f)
n=length(f);
ft=-fft([0,f']',2*n+2);
lf=imag(ft(2:n+1));
```

To prepare this problem for `nsoli.m` we must first consolidate h and c into a single vector $x = (h^T, c^T)^T$. The function `oz.m` does this, organizing the computation as was done in [47]. We also use global variables to avoid repeated computations of the potential u in (3.10). The MATLAB code `ozdemo.m` solves this problem on a 201-point mesh, plots h and c as functions of r, and compares the cost of a few strategies for computing the forcing term.

Here is the part of `ozdemo.m` that produces the graph of the solution in Figure 3.2.

```
% OZDEMO
% This program creates the OZ example in Chapter 3.
% [H,C]=OZDEMO returns the solution on a grid with a mesh
% spacing of 1/256.
%
function [h,c]=ozdemo
global L U rho
n=257;
epsilon=.1; sigma=2.0; rho=.2; beta=10; L=9;
dx=L/(n-1); r=0:dx:L; r=r';
%
% Compute the potential and store it in a global variable.
%
```

```
U=elj(r,sigma,epsilon,beta);
%
tol=[1.d-8,1.d-8];
x=zeros(2*n,1);
parms=[40,80,-.1];
[sol, it_hist, ierr] = nsoli(x,'oz',tol);
%
% Unpack h and c.
%
h=sol(1:n); c=sol(n+1:2*n);
%
% Plot the solution.
%
subplot(1,2,1); plot(r,h,'-');
ylabel('h','Rotation',1); xlabel('r');
subplot(1,2,2); plot(r,c,'-');
ylabel('c','Rotation',1); xlabel('r');
```

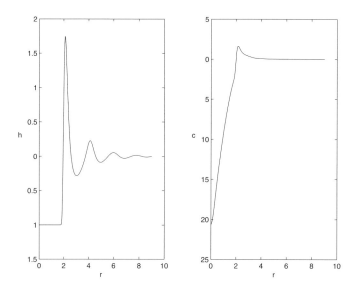

Figure 3.2. *Solution of the OZ equations.*

It is easy to compare methods for computing the forcing term with nsoli.m. In Figure 3.3, also produced with ozdemo.m, we compare the default strategy (3.6) with $\eta = 0.1$. For both computations the tolerances were $\tau_a = \tau_r = 10^{-8}$, much smaller than is needed for a mesh this coarse and used only to illustrate the differences between the choices of the forcing term. For this example, the default choice of η is best if the goal is very small residuals, but the choice $\eta = 0.1$ is superior for realistic values of the tolerances.

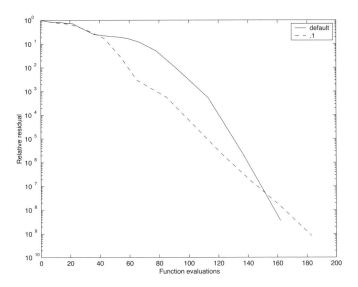

Figure 3.3. *Nonlinear residual versus calls to F.*

3.6.3 Convection-Diffusion Equation

This example is taken from [42] and shows how to incorporate left and right preconditioning into F. The problem is a semilinear (i.e., linear in the highest order derivative) convection-diffusion equation

$$-\nabla^2 u + 20u(u_x + u_y) = f \qquad (3.15)$$

with homogeneous Dirichlet boundary conditions on the unit square $(0,1) \times (0,1)$. Here ∇^2 is the Laplacian operator

$$\nabla^2 = \frac{\partial^2}{\partial^2 x} + \frac{\partial^2}{\partial^2 y}.$$

f has been constructed so that the exact solution is the discretization of

$$10xy(1-x)(1-y)\exp(x^{4.5}).$$

We discretized on a uniform mesh with 31 interior grid points in each direction using centered differences and terminated the iterations consistently with the second-order accuracy of the difference scheme by setting

$$\tau_a = \tau_r = h^2/10.$$

The physical grid is two-dimensional, but solvers expect one-dimensional vectors. MATLAB makes it easy to alternate between a two-dimensional u (padded with the zero boundary conditions), where one applies the differential operators, and the one-dimensional vector that the solvers require. All of this was done within the matrix-free difference operators `dxmf.m` ($\partial/\partial x$), `dymf.m` ($\partial/\partial y$), and `lapmf.f` (Laplacian). As an example, here is the source of `dxmf.m`.

```
function dxu = dxmf(u)
% DXMF Matrix-free partial derivative wrt x;
% homogeneous Dirichlet BC
%
n2=length(u);
n=sqrt(n2);
h=1/(n+1);
%
% Turn u into a 2-D array with the BCs built in.
%
uu=zeros(n+2,n+2);
vv=zeros(n,n);
vv(:)=u;
uu(2:n+1,2:n+1)=vv;
%
% Compute the partial derivative.
%
dxuu=zeros(n,n);
dxuu=uu(3:n+2,2:n+1)-uu(1:n,2:n+1);
%
% Divide by 2*h.
%
dxu=.5*dxuu(:)/h;
i
```

We can exploit the regular grid by using a **fast Poisson solver** M as a preconditioner. Our solver `fish2d.m` uses the MATLAB FFT to solve the discrete form of

$$-\nabla^2 g = u$$

with homogeneous Dirichlet boundary conditions to return $g = Mu$. We apply the preconditioner from the left to (3.15) to obtain the preconditioned equation

$$u + 20M(u(u_x + u_y)) = Mf.$$

`pdeleft.m` is the MATLAB code for the nonlinear residual. Notice that the preconditioner is applied to the low-order term. There is no need to apply `fish2d.m` to `lampmf(u)` simply to recover u.

```
function w=pdeleft(u)
% PDELEFT W=PDELEFT(U) is the nonlinear residual of the left-
%         preconditioned pde example with C=20.
%
global rhsf prhsf
%
% Compute the low-order nonlinear term.
%
```

```
v=20*u.*(dxmf(u)+dymf(u));
%
% Apply fish2d to the entire pde.
%
w=u+fish2d(v)-prhsf;
```

The preconditioned right side

$$Mf = u^* + 20M(u^*(u_x^* + u_y^*)) \tag{3.16}$$

is stored as the global variable prhsf in pdeldemo.m, which calls the solver.

For the right-preconditioned problem, we set $u = Mw$ and solve

$$w + 20(u(u_x + u_y)) = f.$$

The MATLAB code for this is pderight.m. Recall that the residual has a different meaning than for the left-preconditioned problem, so it isn't completely valid to compare the left- and right-preconditioned iterations. It does make sense to compare the choices for linear solvers and forcing terms, however.

Preconditioning a semilinear boundary value problem with an exact inverse for the high-order term, as we do here, is optimal in the sense that the convergence speed of the linear iteration will be independent of the mesh spacing [50]. Multigrid or domain decomposition preconditioners [9,72] also do this, but are more complicated to implement.

One can examine the performance for the three linear solvers and find, as before, that, while the number of nonlinear iterations is roughly the same, the number of function evaluations is lower for GMRES. Were this a larger problem, say, in three space dimensions, the storage for full GMRES could well be unavailable and the low-storage solvers could be the only choices.

The Armijo rule made a difference for the right-preconditioned problem. With right preconditioning, the step length was reduced once at the first nonlinear iteration for all three choices of linear solver and, for BiCGSTAB and TFQMR, once again on the second nonlinear iteration.

3.6.4 Time-Dependent Convection-Diffusion Equation

This example is a time-dependent form of the equation in section 3.6.3. We will use the implicit Euler method to integrate the nonlinear parabolic initial boundary value problem

$$u_t = \nabla^2 u - 20u(u_x + u_y) - f \tag{3.17}$$

in time for $0 < t < 1$. As in section 3.6.3, we impose homogeneous Dirichlet boundary conditions on the unit square $(0, 1) \times (0, 1)$. The function $f(x, y)$ is the same as in section 3.6.3, so the solution of the steady-state (time-independent) problem is

$$u_{steady}(x, y) = 10xy(1 - x)(1 - y) \exp(x^{4.5}).$$

We expect the solution $u(x, t)$ of (3.17) to converge to u_{steady} as $t \to \infty$. After discretization in space, the problem becomes a large system of ordinary differential equations. This system is stiff, so implicit methods are necessary if we want to avoid unreasonably small time steps.

We follow the procedure from section 2.7.5 to prepare the nonlinear systems that must be solved for each time step. First, we discretize in space with centered differences to obtain a system of ordinary differential equations, which we write as

$$u_t = N(u),\ u(0) = u^0. \tag{3.18}$$

The nonlinear equation that must be solved for the implicit Euler method with a fixed time step of δ_t is

$$u^{n+1} - u^n = \delta_t N(u^{n+1}). \tag{3.19}$$

To precondition from the left with the fast Poisson solver `fish2d.m`, one solves, at each time step, the nonlinear equation

$$F(U) = M(U - u^n - \delta_t N(U)) = 0,$$

where M represents the application of the `fish2d.m` solver.

The code `pdetime.m` for the solve is shorter than the explanation above. Because the nonlinear residual F has been constructed, integration in time proceeds just as it did in section 2.7.5. The integration code is `pdetimedemo.m`, which uses a 63×63 spatial mesh and, a time step of 0.1, solves linear systems with GMRES, and, at the end, compares the result at $t = 1$ with the steady-state solution.

3.7 Projects

3.7.1 Krylov Methods and the Forcing Term

Compare the performance of the three Krylov methods and various choices of the forcing term for the H-equation, the OZ equations, and the convection-diffusion equation. Make the comparison in terms of computing time, number of function evaluations needed to reach a given tolerance, and storage requirements. If GMRES is limited to the storage that BiCGSTAB or TFQMR needs, how well does it perform? Do all choices of the forcing term lead to the same root?

3.7.2 Left and Right Preconditioning

Use the `pdeldemo.m` and `pderdemo.m` codes in the directory for this chapter to examine the quality of the Poisson solver preconditioner. Modify the codes to refine the mesh and see if the performance of the preconditioner degrades as the mesh is refined. Compare the accuracy of the results.

3.7.3 Two-Point Boundary Value Problem

Try to solve the boundary value problem from section 2.7.4 with `nsoli.m`. You'll need a preconditioner to get good performance. Try using a factorization of $F'(x_0)$ to build one. Would an incomplete factorization (like `luinc` from MATLAB) work?

3.7.4 Making a Movie

The code `ozmovie.m` in the directory for this chapter solves the OZ equations and makes a movie of the iterations for h. Use this code as a template to make movies of solutions, steps, and nonlinear residuals for some of the other examples. This is especially interesting for the differential equation problems.

3.8 Source Code for nsoli.m

```
1   function [sol, it_hist, ierr, x_hist] = nsoli(x,f,tol, parms)
2   % NSOLI  Newton-Krylov solver, globally convergent
3   %        solver for f(x) = 0
4   %
5   % Inexact Newton-Armijo iteration
6   %
7   % Eisenstat-Walker forcing term
8   %
9   % Parabolic line search via three-point interpolation
10  %
11  % C. T. Kelley, April 1, 2003
12  %
13  % This code comes with no guarantee or warranty of any kind.
14  %
15  % function [sol, it_hist, ierr, x_hist] = nsoli(x,f,tol,parms)
16  %
17  % inputs:
18  %       initial iterate = x
19  %       function = f
20  %       tol = [atol, rtol] relative/absolute
21  %              error tolerances for the nonlinear iteration
22  %       parms = [maxit, maxitl, etamax, lmeth, restart_limit]
23  %       maxit = maximum number of nonlinear iterations
24  %              default = 40
25  %       maxitl = maximum number of inner iterations before restart
26  %              in GMRES(m), m = maxitl
27  %              default = 40
28  %
29  %              For iterative methods other than GMRES(m) maxitl
30  %              is the upper bound on linear iterations.
31  %
32  %       |etamax| = Maximum error tolerance for residual in inner
33  %              iteration. The inner iteration terminates
34  %              when the relative linear residual is
35  %              smaller than eta*| F(x_c) |. eta is determined
36  %              by the modified Eisenstat-Walker formula if etamax > 0.
37  %              If etamax < 0, then eta = |etamax| for the entire
38  %              iteration.
39  %              default: etamax = 0.9
40  %
41  %       lmeth = choice of linear iterative method
42  %              1 (GMRES), 2 GMRES(m),
43  %              3 (BICGSTAB), 4 (TFQMR)
44  %              default = 1 (GMRES, no restarts)
45  %
46  %       restart_limit = max number of restarts for GMRES if
47  %              lmeth = 2
48  %              default = 20
49  %
50  % output:
51  %       sol = solution
52  %       it_hist(maxit,3) = l2-norms of nonlinear residuals
53  %              for the iteration, number of function evaluations,
54  %              and number of step-length reductions
55  %       ierr = 0 upon successful termination
56  %       ierr = 1 if after maxit iterations
57  %              the termination criterion is not satisfied
58  %       ierr = 2 failure in the line search. The iteration
59  %              is terminated if too many step-length reductions
60  %              are taken.
61  %
62  %       x_hist = matrix of the entire iteration history.
63  %              The columns are the nonlinear iterates. This
64  %              is useful for making movies, for example, but
65  %              can consume way too much storage. This is an
66  %              OPTIONAL argument. Storage is only allocated
67  %              if x_hist is in the output argument list.
68  %
69  %
70  % internal parameters:
71  %       debug = turns on/off iteration statistics display as
72  %              the iteration progresses
73  %
74  %       alpha = 1.d-4, parameter to measure sufficient decrease
75  %
76  %       sigma0 = 0.1, sigma1 = 0.5, safeguarding bounds for the line search
77  %
78  %       maxarm = 20, maximum number of step-length reductions before
79  %              failure is reported
80  %
81  %
82  %
83  % Set the debug parameter; 1 turns display on, otherwise off.
84  %
85  debug = 0;
86  %
87  % Set internal parameters.
88  %
89  alpha = 1.d-4; sigma0 = .1; sigma1 = .5; maxarm = 20; gamma = .9;
90  %
91  % Initialize it_hist, ierr, x_hist and set the default values of
92  % those iteration parameters that are optional inputs.
93  %
94  ierr = 0; maxit = 40; lmaxit = 40; etamax = .9; it_histx = zeros(maxit,3);
95  lmeth = 1; restart_limit = 20;
96  if nargout == 4, x_hist = x; end
97  %
98  % Initialize parameters for the iterative methods.
99  
```

```
101 % Check for optional inputs.
102 %
103 gmparms = [abs(etamax), lmaxit];
104 if nargin == 4
105     maxit = parms(1); lmaxit = parms(2); etamax = parms(3);
106     it_histx = zeros(maxit,3);
107     gmparms = [abs(etamax), lmaxit];
108     if length(parms)>= 4
109         lmeth = parms(4);
110     end
111     if length(parms) == 5
112         gmparms = [abs(etamax), lmaxit, parms(5), 1];
113     end
114 end
115 %
116 rtol = tol(2); atol = tol(1); n = length(x); fnrm = 1; itc = 0;
117 % Evaluate f at the initial iterate, and
118 % compute the stop tolerance.
119 %
120 %
121 f0 = feval(f,x);
122 fnrm = norm(f0);
123 it_histx(itc+1,1) = fnrm; it_histx(itc+1,2) = 0; it_histx(itc+1,3) = 0;
124 fnrmo = 1;
125 stop_tol = atol + rtol*fnrm;
126 outstat(itc+1, :) = [itc fnrm 0 0 0];
127 %
128 % main iteration loop
129 %
130 while(fnrm > stop_tol & itc < maxit)
131 %
132 % Keep track of the ratio (rat = fnrm/frnmo)
133 % of successive residual norms and
134 % the iteration counter (itc).
135 %
136     rat = fnrm/fnrmo;
137     fnrmo = fnrm;
138     itc = itc+1;
139     [step, errstep, inner_it_count,inner_f_evals] = ...
140         dkrylov(f0, f, x, gmparms, lmeth);
141 %
142 % The line search starts here.
143 %
144 %
145     xold = x;
146     lambda = 1; lamm = 1; lamc = lambda; iarm = 0;
147     xt = x + lambda*step;
148     ft = feval(f,xt);
149     nft = norm(ft); nf0 = norm(f0); ff0 = nf0*nf0; ffc = nft*nft; ffm = nft*nft;
150     while nft >= (1 - alpha*lambda) * nf0;
151 %
152 % Apply the three-point parabolic model.
153 %
154     if iarm == 0
155         lambda = sigma1*lambda;
156     else
157         lambda = parab3p(lamc, lamm, ff0, ffc, ffm);
158     end
159 % Update x; keep the books on lambda.
160 %
161 %
162     xt = x+lambda*step;
163     lamm = lamc;
164     lamc = lambda;
165 %
166 % Keep the books on the function norms.
167 %
168     ft = feval(f,xt);
169     nft = norm(ft);
170     ffm = ffc;
171     ffc = nft*nft;
172     iarm = iarm+1;
173     if iarm > maxarm
174         disp('Armijo failure, too many reductions');
175         ierr = 2;
176         disp(outstat)
177         it_hist = it_histx(1:itc+1,:);
178         if nargout == 4, x_hist = [x_hist,xt]; end
179         sol = xold;
180         return;
181     end
182     end
183     x = xt;
184     f0 = ft;
185 % end of line search
186 %
187 %
188     if nargout == 4, x_hist = [x_hist,x]; end
189     fnrm = norm(f0);
190     it_histx(itc+1,1) = fnrm;
191 %
192 % How many function evaluations did this iteration require?
193 %
194     it_histx(itc+1,2) = it_histx(itc,2)+inner_f_evals+iarm+1;
195     if itc == 1, it_histx(itc+1,2) = it_histx(itc+1,2)+1; end;
196     it_histx(itc+1,3) = iarm;
197 %
198     rat = fnrm/fnrmo;
199 %
200 % Adjust eta as per Eisenstat-Walker
201 %
202     if etamax > 0
```

```
203         etaold = gmparms(1);
204         etanew = gamma*rat*rat;
205         if gamma*etaold*etaold > .1
206             etanew = max(etanew,gamma*etaold*etaold);
207         end
208         gmparms(1) = min([etanew,etamax]);
209         gmparms(1) = max(gmparms(1),.5*stop_tol/fnrm);
210     end
211 %
212     outstat(itc+1, :) = [itc fnrm inner_it_count rat iarm];
213 %
214 end
215 sol = x;
216 it_hist = it_histx(1:itc+1,:);
217 if debug == 1
218     disp(outstat)
219     it_hist = it_histx(1:itc+1,:);
220 end
221 %
222 % On failure, set the error flag.
223 %
224 if fnrm > stop_tol, ierr = 1; end
225 %
226 %
227 function lambdap = parab3p(lambdac, lambdam, ff0, ffc, ffm)
228 % Apply three-point safeguarded parabolic model for a line search.
229 %
230 % C. T. Kelley, April 1, 2003
231 %
232 % This code comes with no guarantee or warranty of any kind.
233 %
234 function lambdap = parab3p(lambdac, lambdam, ff0, ffc, ffm)
235 %
236 % input:
237 %       lambdac = current step length
238 %       lambdam = previous step length
239 %       ff0 = value of || F(x_c) ||^2
240 %       ffc = value of || F(x_c + lambdac d) ||^2
241 %       ffm = value of || F(x_c + lambdam d) ||^2
242 %
243 % output:
244 %       lambdap = new value of lambda given parabolic model
245 %
246 % internal parameters:
247 %       sigma0 = 0.1, sigma1 = 0.5, safeguarding bounds for the line search
248 %
249 %
250 %
251 % Set internal parameters.
252 %
253 sigma0 = .1; sigma1 = .5;

254 %
255 % Compute coefficients of interpolation polynomial.
256 %
257 % p(lambda) = ff0 + (c1 lambda + c2 lambda^2)/d1
258 %
259 % d1 = (lambdac - lambdam)*lambdam*lambdam < 0
260 %      so if c2 > 0 we have negative curvature and default to
261 %      lambda = sigma1 * lambda.
262 %
263 c2 = lambdam*(ffc-ff0)-lambdac*(ffm-ff0);
264 if c2 >= 0
265     lambdap = sigma1*lambdac; return
266 end
267 c1 = lambdac*lambdac*(ffm-ff0)-lambdam*lambdam*(ffc-ff0);
268 lambdap = -c1*.5/c2;
269 if lambdap < sigma0*lambdac, lambdap = sigma0*lambdac; end
270 if lambdap > sigma1*lambdac, lambdap = sigma1*lambdac; end
271 %
272 %
273 %
274 function [step, errstep, total_iters, f_evals] = ...
275           dkrylov(f0, f, x, params, lmeth)
276 % Krylov linear equation solver for use in nsoli
277 %
278 % C. T. Kelley, April 1, 2003
279 %
280 %
281 % This code comes with no guarantee or warranty of any kind.
282 %
283 function [step, errstep, total_iters, f_evals]
284                       = dkrylov(f0, f, x, params, lmeth)
285 %
286 %
287 % Input:  f0 = function at current point
288 %         f = nonlinear function
289 %             The format for f is function fx = f(x).
290 %             Note that for Newton-GMRES we incorporate any
291 %             preconditioning into the function routine.
292 %         x = current point
293 %         params = vector to control iteration
294 %             params(1) = relative residual reduction factor
295 %             params(2) = max number of iterations
296 %             params(3) = max number of restarts for GMRES(m)
297 %             params(4) (Optional) = reorthogonalization method in GMRES
298 %                   1 -- Brown/Hindmarsh condition (default)
299 %                   2 -- Never reorthogonalize (not recommended)
300 %                   3 -- Always reorthogonalize (not cheap!)
301 %
302 %         lmeth = method choice
303 %             1 GMRES without restarts (default)
304 %             2 GMRES(m) , m = params(2) and the maximum number
```

```matlab
305 %            of restarts is params(3).
306 %        3 BiCGSTAB
307 %        4 TFQMR
308 %
309 % Output: x = solution
310 %         errstep = vector of residual norms for the history of
311 %                   the iteration
312 %         total_iters = number of iterations
313 %
314 %
315 %
316 % initialization
317 %
318 lmaxit = params(2);
319 restart_limit = 20;
320 if length(params) >= 3
321     restart_limit = params(3);
322 end
323 if lmeth == 1, restart_limit = 0; end
324 if length(params) == 3
325 % default reorthogonalization
326 %
327     gmparms = [params(1), params(2), 1];
328 elseif length(params) == 4
329 % reorthogonalization method is params(4)
330 %
331     gmparms = [params(1), params(2), params(4)];
332 else
333     gmparms = [params(1), params(2), params(2)];
334 end
335 %
336 % linear iterative methods
337 %
338 if lmeth == 1 | lmeth == 2   % GMRES or GMRES(m)
339 % Compute the step using a GMRES routine especially designed
340 % for this purpose.
341 %
342     [step, errstep, total_iters] = dgmres(f0, f, x, gmparms);
343     kinn = 0;
344 %
345 % Restart at most restart_limit times.
346 %
347     while total_iters == lmaxit & ...
348         errstep(total_iters) > gmparms(1)*norm(f0) & ...
349         kinn < restart_limit
350         kinn = kinn+1;
351         [step, errstep, total_iters] = dgmres(f0, f, x, gmparms,step);
352
353
354
355     end
356     total_iters = total_iters+kinn*lmaxit;
357     f_evals = total_iters+kinn;
358 %
359 % BiCGSTAB
360 %
361 elseif lmeth == 3
362     [step, errstep, total_iters] = dcgstab(f0, f, x, gmparms);
363     f_evals = 2*total_iters;
364 %
365 % TFQMR
366 %
367 elseif lmeth == 4
368     [step, errstep, total_iters] = dtfqmr(f0, f, x, gmparms);
369     f_evals = 2*total_iters;
370 else
371     error('lmeth error in fdkrylov')
372 end
373 %
374 %
375 function z = dirder(x,w,f,f0)
376 % Finite difference directional derivative
377 % Approximate f'(x) w.
378 %
379 % C. T. Kelley, April 1, 2003
380 %
381 % This code comes with no guarantee or warranty of any kind.
382 %
383 % function z = dirder(x,w,f,f0)
384 %
385 % inputs:
386 %           x, w = point and direction
387 %           f = function
388 %           f0 = f(x), in nonlinear iterations.
389 %                f(x) has usually been computed
390 %                before the call to dirder.
391 %
392 % Use a hardwired difference increment.
393 %
394 epsnew = 1.d-7;
395 %
396 n = length(x);
397 %
398 % Scale the step.
399 %
400 if norm(w) == 0
401     z = zeros(n,1);
402     return
403 end
404 %
405 %
406 %
```

80

```
407 % Now scale the difference increment.
408 %
409 xs=(x'*w)/norm(w);
410 if xs ~= 0.d0
411     epsnew=epsnew*max(abs(xs),1.d0)*sign(xs);
412 end
413 epsnew=epsnew/norm(w);
414 %
415 % del and f1 could share the same space if storage
416 % is more important than clarity.
417 %
418 del = x+epsnew*w;
419 f1 = feval(f,del);
420 z = (f1 - f0)/epsnew;
421 %
422 %
423 function [x, error, total_iters] = dgmres(f0, f, xc, params, xinit)
424 % GMRES linear equation solver for use in Newton-GMRES solver
425 %
426 % C. T. Kelley, April 1, 2003
427 %
428 % This code comes with no guarantee or warranty of any kind.
429 %
430 function [x, error, total_iters] = dgmres(f0, f, xc, params, xinit)
431 %
432 %
433 % Input:  f0 = function at current point
434 %         f = nonlinear function
435 %            The format for f is function fx = f(x).
436 %            Note that for Newton-GMRES we incorporate any
437 %            preconditioning into the function routine.
438 %         xc = current point
439 %         params = vector to control iteration
440 %            params(1) = relative residual reduction factor
441 %            params(2) = max number of iterations
442 %            params(3) (Optional) = reorthogonalization method
443 %                1 -- Brown/Hindmarsh condition (default)
444 %                2 -- Never reorthogonalize (not recommended)
445 %                3 -- Always reorthogonalize (not cheap!)
446 %
447 %         xinit = initial iterate. xinit = 0 is the default. This
448 %            is a reasonable choice unless restarted GMRES
449 %            will be used as the linear solver.
450 %
451 % Output: x = solution
452 %         error = vector of residual norms for the history of
453 %            the iteration
454 %         total_iters = number of iterations
455 %
456 % requires givapp.m, dirder.m
457 %
458 %
459 % initialization
460 %
461 errtol = params(1);
462 kmax = params(2);
463 reorth = 1;
464 if length(params) == 3
465     reorth = params(3);
466 end
467 %
468 % The right side of the linear equation for the step is -f0.
469 %
470 b = -f0;
471 n = length(b);
472 %
473 % Use zero vector as initial iterate for Newton step unless
474 % the calling routine has a better idea (useful for GMRES(m)).
475 %
476 x = zeros(n,1);
477 r = b;
478 if nargin == 5
479     x = xinit;
480     r = -dirder(xc, x, f, f0)-f0;
481 end
482 %
483 %
484 h = zeros(kmax);
485 v = zeros(n,kmax);
486 c = zeros(kmax+1,1);
487 s = zeros(kmax+1,1);
488 rho = norm(r);
489 g = rho*eye(kmax+1,1);
490 errtol = errtol*norm(b);
491 error = [];
492 %
493 % Test for termination on entry.
494 %
495 error = [error,rho];
496 total_iters = 0;
497 if(rho < errtol)
498     disp('early termination')
499     return
500 end
501 %
502 %
503 v(:,1) = r/rho;
504 beta = rho;
505 k = 0;
506 %
507 % GMRES iteration
508 %
```

```
509 while((rho > errtol) & (k < kmax))
510     k = k+1;
511 %
512 %   Call directional derivative function.
513 %
514     v(:,k+1) = dirder(xc, v(:,k), f, f0);
515     normav = norm(v(:,k+1));
516 %
517 %   Modified Gram-Schmidt
518 %
519     for j = 1:k
520         h(j,k) = v(:,j)'*v(:,k+1);
521         v(:,k+1) = v(:,k+1)-h(j,k)*v(:,j);
522     end
523     h(k+1,k) = norm(v(:,k+1));
524     normav2 = h(k+1,k);
525 %
526 %   Reorthogonalize?
527 %
528 if  (reorth == 1 & normav + .001*normav2 == normav) | reorth == 3
529     for j = 1:k
530         hr = v(:,j)'*v(:,k+1);
531         h(j,k) = h(j,k)+hr;
532         v(:,k+1) = v(:,k+1)-hr*v(:,j);
533     end
534     h(k+1,k) = norm(v(:,k+1));
535 end
536 %
537 %   Watch out for happy breakdown.
538 %
539 if(h(k+1,k) ~= 0)
540     v(:,k+1) = v(:,k+1)/h(k+1,k);
541 end
542 %
543 %   Form and store the information for the new Givens rotation.
544 %
545 if k > 1
546     h(1:k,k) = givapp(c(1:k-1),s(1:k-1),h(1:k,k),k-1);
547 end
548 %
549 %   Don't divide by zero if solution has been found.
550 %
551     nu = norm(h(k:k+1,k));
552     if nu ~= 0
553 %        c(k) = h(k,k)/nu;
554         c(k) = conj(h(k,k)/nu);
555         s(k) = -h(k+1,k)/nu;
556         h(k,k) = c(k)*h(k,k)-s(k)*h(k+1,k);
557         h(k+1,k) = 0;
558         g(k:k+1) = givapp(c(k),s(k),g(k:k+1),1);
559     end
560 %
561 %   Update the residual norm.
562 %
563     rho = abs(g(k+1));
564     error = [error,rho];
565 %
566 %   end of the main while loop
567 %
568 end
569 %
570 %   At this point either k > kmax or rho < errtol.
571 %   It's time to compute x and leave.
572 %
573 y = h(1:k,1:k)\g(1:k);
574 total_iters = k;
575 x = x + v(1:n,1:k)*y;
576 %
577 %
578 function vrot = givapp(c,s,vin,k)
579 %  Apply a sequence of k Givens rotations, used within GMRES codes.
580 %
581 %  C. T. Kelley, April 1, 2003
582 %
583 %  This code comes with no guarantee or warranty of any kind.
584 %
585 % function vrot = givapp(c, s, vin, k)
586 %
587 vrot = vin;
588 for i = 1:k
589     w1 = c(i)*vrot(i)-s(i)*vrot(i+1);
590 %
591 %  Here's a modest change that makes the code work in complex
592 %  arithmetic. Thanks to Howard Elman for this.
593 %
594 %    w2 = s(i)*vrot(i)+c(i)*vrot(i+1);
595     w2 = s(i)*vrot(i)+conj(c(i))*vrot(i+1);
596     vrot(i:i+1) = [w1,w2];
597 end
598 %
599 %
600 function [x, error, total_iters] = ...
601                    dcgstab(f0, f, xc, params, xinit)
602 % Forward difference BiCGSTAB solver for use in nsoli
603 %
604 % C. T. Kelley, April 1, 2003
605 %
606 % This code comes with no guarantee or warranty of any kind.
607 %
608 % function [x, error, total_iters]
609 %               = dcgstab(f0, f, xc, params, xinit)
610 %
```

82

```
611 % Input:    f0 = function at current point
612 %           f = nonlinear function
613 %               The format for f is function fx = f(x).
614 %               Note that for Newton-GMRES we incorporate any
615 %               preconditioning into the function routine.
616 %           xc = current point
617 %           params = two-dimensional vector to control iteration
618 %               params(1) = relative residual reduction factor
619 %               params(2) = max number of iterations
620 %
621 %           xinit = initial iterate. xinit = 0 is the default. This
622 %               is a reasonable choice unless restarts are needed.
623 %
624 %
625 % Output: x = solution
626 %           error = vector of residual norms for the history of
627 %               the iteration
628 %           total_iters = number of iterations
629 %
630 % Requires: dirder.m
631 %
632 %
633 % initialization
634 %
635 %
636 b = -f0;
637 n = length(b); errtol = params(1)*norm(b); kmax = params(2); error = [];
638 rho = zeros(kmax+1,1);
639 %
640 % Use zero vector as initial iterate for Newton step unless
641 % the calling routine has a better idea (useful for GMRES(m)).
642 %
643 x = zeros(n,1);
644 r = b;
645 if nargin == 5
646     x = xinit;
647     r = -dirder(xc, x, f, f0)-f0;
648 end
649 %
650 hatr0 = r;
651 k = 0; rho(1) = 1; alpha = 1; omega = 1;
652 v = zeros(n,1); p = zeros(n,1); rho(2) = hatr0'*r;
653 zeta = norm(r); error = [error,zeta];
654 %
655 % BiCGSTAB iteration
656 %
657 while((zeta > errtol) & (k < kmax))
658     k = k+1;
659     if omega == 0
660         error('BiCGSTAB breakdown, omega = 0');
661     end
662     beta = (rho(k+1)/rho(k))*(alpha/omega);
663     p = r+beta*(p - omega*v);
664     v = dirder(xc,p,f,f0);
665     tau = hatr0'*v;
666     if tau == 0
667         error('BiCGSTAB breakdown, tau = 0');
668     end
669     alpha = rho(k+1)/tau;
670     s = r-alpha*v;
671     t = dirder(xc,s,f,f0);
672     tau = t'*t;
673     if tau == 0
674         error('BiCGSTAB breakdown, t = 0');
675     end
676     omega = t'*s/tau;
677     rho(k+2) = -omega*(hatr0'*t);
678     x = x+alpha*p+omega*s;
679     r = s-omega*t;
680     zeta = norm(r);
681     total_iters = k;
682     error = [error, zeta];
683 end
684 %
685 %
686 %
687 function [x, error, total_iters] = ...
688                 dtfqmr(f0, f, xc, params, xinit)
689 % Forward difference TFQMR solver for use in nsoli
690 %
691 % C. T. Kelley, April 1, 2003
692 %
693 % This code comes with no guarantee or warranty of any kind.
694 %
695 function [x, error, total_iters]
696                 = dtfqmr(f0, f, xc, params, xinit)
697 %
698 %
699 %
700 % Input:    f0 = function at current point
701 %           f = nonlinear function
702 %               The format for f is function fx = f(x).
703 %               Note that for Newton-GMRES we incorporate any
704 %               preconditioning into the function routine.
705 %           xc = current point
706 %           params = vector to control iteration
707 %               params(1) = relative residual reduction factor
708 %               params(2) = max number of iterations
709 %
710 %           xinit = initial iterate. xinit = 0 is the default. This
711 %               is a reasonable choice unless restarts are needed.
712 %
```

```
713 %
714 % Output: x = solution
715 %         error = vector of residual norms for the history of
716 %               the iteration
717 %         total_iters = number of iterations
718 %
719 % Requires: dirder.m
720 %
721
722 % initialization
723 %
724 %
725 b = -f0;
726 n = length(b); errtol = params(1)*norm(b); kmax = params(2); error = [];
727 x = zeros(n,1);
728 r = b;
729 if nargin == 5
730     x = xinit;
731     r = -dirder(xc, x, f, f0)-f0;
732 end
733
734 u = zeros(n,2); y = zeros(n,2); w = r; y(:,1) = r;
735 k = 0; d = zeros(n,1);
736 v = dirder(xc, y(:,1),f,f0);
737 u(:,1) = v;
738 theta = 0; eta = 0; tau = norm(r); error = [error,tau];
739 rho = tau*tau;
740 %
741 % TFQMR iteration
742 %
743 while( k < kmax)
744     k = k+1;
745     sigma = r'*v;
746 %
747     if sigma == 0
748         error('TFQMR breakdown, sigma = 0')
749     end
750
751     alpha = rho/sigma;
752 %
753 %
754 %
755     for j = 1:2
756 %
757 % Compute y2 and u2 only if you have to.
758 %
759         if j == 2
760             y(:,2) = y(:,1)-alpha*v;
761             u(:,2) = dirder(xc, y(:,2),f,f0);
762         end
763         m = 2*k-2+j;
764         w = w-alpha*u(:,j);
765         d = y(:,j)+(theta*theta*eta/alpha)*d;
766         theta = norm(w)/tau; c = 1/sqrt(1+theta*theta);
767         tau = tau*theta*c; eta = c*c*alpha;
768         x = x+eta*d;
769 %
770 % Try to terminate the iteration at each pass through the loop.
771 %
772         if tau*sqrt(m+1) <=   errtol
773             error = [error, tau];
774             total_iters = k;
775             return
776         end
777     end
778 %
779 %
780 %
781     if rho == 0
782         error('TFQMR breakdown, rho = 0')
783     end
784 %
785     rhon = r'*w; beta = rhon/rho; rho = rhon;
786     y(:,1) = w + beta*y(:,2);
787     u(:,1) = dirder(xc, y(:,1),f,f0);
788     v = u(:,1)+beta*(u(:,2)+beta*v);
789     error = [error,tau];
790     total_iters = k;
791 end
792 %
793 %
```

Chapter 4

Broyden's Method

Broyden's method [14] approximates the Newton direction by using an approximation of the Jacobian (or its inverse), which is updated as the nonlinear iteration progresses. The cost of this updating in the modern implementation we advocate here is one vector for each nonlinear iteration. Contrast this cost to Newton–GMRES, where the storage is accumulated during a linear iteration. For a problem where the initial iterate is far from a solution and the number of nonlinear iterations will be large, this is a significant disadvantage for Broyden's method. Broyden's method, like the secant method for scalar equations, does not guarantee that the approximate Newton direction will be a descent direction for $\|F\|$ and therefore a line search may fail. For these reasons, the Newton–Krylov methods are now (2003) used more frequently than Broyden's method. Having said that, when the initial iterate is near the solution, Broyden's method can perform very well.

Broyden's method usually requires preconditioning to perform well, so the decisions you will make are the same as those for a Newton–Krylov method.

Broyden's method is the simplest of the **quasi-Newton** methods. These methods are extensions of the secant method to several variables. Recall that the secant method approximates $f'(x_n)$ with

$$b_n = \frac{f(x_n) - f(x_{n-1})}{x_n - x_{n-1}} \tag{4.1}$$

and then takes the step

$$x_{n+1} = x_n - b_n^{-1} f(x_n).$$

One way to mimic this in higher dimensions is to carry an approximation to the Jacobian along with the approximation to x^* and update the approximate Jacobian as the iteration progresses. The formula for b_n will not do, because one can't divide by a vector. However, one can ask that B_n, the current approximation to $F'(x_n)$, satisfy the secant equation

$$B_n(x_n - x_{n-1}) = F(x_n) - F(x_{n-1}). \tag{4.2}$$

For scalar equations, (4.2) and (4.1) are equivalent. For equations in more than one variable, (4.1) is meaningless, so a wide variety of methods that satisfy the secant equation have been designed to preserve such properties of the Jacobian as the sparsity pattern or symmetry [24, 42, 43].

In the case of Broyden's method, if x_n and B_n are the current approximate solution and Jacobian, respectively, then

$$x_{n+1} = x_n - \lambda_n B_n^{-1} F(x_n), \tag{4.3}$$

where λ_n is the step length for the Broyden direction

$$d_n = -B_n^{-1} F(x_n).$$

After the computation of x_{n+1}, B_n is **updated** to form B_{n+1} using the Broyden update

$$B_{n+1} = B_n + \frac{(y - B_n s)s^T}{s^T s}. \tag{4.4}$$

In (4.4), $y = F(x_{n+1}) - F(x_n)$ and

$$s = x_{n+1} - x_n = \lambda_n d_n.$$

4.1   Convergence Theory

The convergence theory for Broyden's method is only local and, therefore, less satisfactory than that for the Newton and Newton–Krylov methods. The line search cannot be proved to compensate for a poor initial iterate. Theorem 4.1 is all there is.

Theorem 4.1.   *Let the standard assumptions hold. Then there are δ and δ_B such that, if*

$$\|x_0 - x^*\| < \delta \ \text{ and } \ \|B_0 - F'(x^*)\| < \delta_B,$$

then the Broyden sequence for the data (F, x_0, B_0) exists and $x_n \to x^*$ q-superlinearly; i.e.,

$$\lim_{n \to \infty} \frac{\|e_{n+1}\|}{\|e_n\|} = 0.$$

4.2   An Algorithmic Sketch

Most implementations of Broyden's method, our code `brsola.m` among them, incorporate a line search. Keep in mind the warning in section 1.7.1! This may not work and a code must be prepared for the line search to fail. The algorithm follows the broad outline of `nsolg`. The data now include an initial approximation B to the Jacobian.

Algorithm 4.1.
broyden_sketch$(x, B, F, \tau_a, \tau_r)$

   Evaluate $F(x)$; $\tau \leftarrow \tau_r |F(x)| + \tau_a$.

   while $\|F(x)\| > \tau$ **do**

      Solve $Bd = -F(x)$.

      Use a line search to compute a step length λ.

      If the line search fails, terminate.

      $s \leftarrow \lambda d$; $y \leftarrow F(x + \lambda d) - F(x)$

      $x \leftarrow x + s$

$$B \leftarrow B + \frac{(y - Bs)s^T}{s^T s}.$$

   end while

The local convergence theory for Broyden's method is completely satisfactory. If the standard assumptions hold and the data x_0 and B_0 are accurate approximations to x^* and $F'(x^*)$, then the convergence is q-superlinear.

There are many ways to obtain a good B_0. If the initial iterate is accurate, $B_0 = F'(x_0)$ is a good choice. Letting B_0 be the highest order term in a discretized elliptic partial differential equation or the noncompact term in an integral equation is another example.

Unlike inexact Newton methods or Newton iterative methods, quasi-Newton methods need only one function evaluation for each nonlinear iteration. The storage requirements for Broyden's method, as we will see, are very similar to those for Newton-GMRES.

4.3 Computing the Broyden Step and Update

One way to solve the equation for the Broyden step is to factor B_n with each iteration. This, of course, eliminates part of the advantage of approximating the Jacobian. One can also factor B_0 and update that factorization (see [24] for one way to do this), but this is also extremely costly. Most quasi-Newton codes update B_n^{-1} as the iteration progresses, using preconditioning to arrange things so that $B_0 = I$.

Left preconditioning works in the following way. Suppose $A \approx F'(x^*)$. Rather than use $B_0 = A$, one could apply Broyden's method to the left-preconditioned problem $A^{-1}F(x) = 0$ and use $B_0 = I$. The two sequences of approximate solutions are exactly the same [42]. If, instead, one uses the right-preconditioned problem $F(A^{-1}x) = 0$, $B_0 = I$ is still a good choice, but the nonlinear iteration will be different.

Keep in mind that one will never compute and store A^{-1}, but rather factor A and store the factors. One then applies this factorization at a cost of $O(N^2)$ floating point operations whenever one wants to compute $A^{-1}F(x)$ or $F(A^{-1}x)$. This will amortize the $O(N^3)$ factorization of A over the entire nonlinear iteration.

The next step is to use the Sherman–Morrison formula [69, 70]. If B is a nonsingular matrix and $u, v \in R^N$, then $B + uv^T$ is nonsingular if and only if

$1 + v^T B^{-1} u \neq 0$. In that case,

$$(B + uv^T)^{-1} = \left(I - \frac{(B^{-1}u)v^T}{1 + v^T B^{-1} u} \right) B^{-1}. \tag{4.5}$$

To apply (4.5) to Broyden's method, we write (4.4) as

$$B_{n+1} = B_n + u_n v_n^T,$$

where

$$u_n = (y_n - B_n s_n)/\|s_n\| \text{ and } v_n = s_n/\|s_n\|.$$

Then, keeping in mind that $B_0 = I$,

$$\begin{aligned} B_n^{-1} &= (I - w_{n-1}v_{n-1}^T)(I - w_{n-2}v_{n-2}^T) \cdots (I - w_0 v_0^T)B_0^{-1} \\ &= \prod_{j=0}^{n-1} (I - w_j v_j^T), \end{aligned} \tag{4.6}$$

where, for $k \geq 0$,

$$w_k = (B_k^{-1} u_k)/(1 + v_k^T B_k^{-1} u_k).$$

So, to apply B_n^{-1} to a vector p, we use (4.6) at a cost of $O(Nn)$ floating point operations and storage of the $2n$ vectors $\{w_k\}_{k=0}^{n-1}$ and $\{s_k\}_{k=0}^{n-1}$. The storage can be halved with a trick [26, 42] using the observation that the search direction satisfies

$$\begin{aligned} d_{n+1} &= -B_{n+1}^{-1} F(x_{n+1}) \\ &= -\left(I - \frac{w_n s_n^T}{\|s_n\|} \right) B_n^{-1} F(x_{n+1}) \\ &= -\frac{\|s_n\|^2 B_n^{-1} F(x_{n+1}) - (1 - \lambda_n)s_n^T B_n^{-1} F(x_{n+1})s_n}{\|s_n\|^2 + \lambda_n s_n^T B_n^{-1} F(x_{n+1})}. \end{aligned}$$

Hence (see [42] for details) one can compute the search direction and update B simultaneously and only have to store one new vector for each nonlinear iteration.

Algorithm **broyden** shows how this is implemented in our Broyden–Armijo MATLAB code **brsola.m**. Keep in mind that we assume that F has been preconditioned and that $B_0 = I$. Note that we also must store the sequence of step lengths.

Algorithm 4.2.
broyden(x, F, τ_a, τ_r)

    Evaluate $F(x)$; $\tau \leftarrow \tau_r |F(x)| + \tau_a$.
    $d \leftarrow -F(x)$; compute λ_0 with a line search.
    Terminate if the line search fails.
    $s_0 \leftarrow \lambda_0 d$; $x \leftarrow x + s$
    $n \leftarrow 0$
    while $\|F(x)\| > \tau$ **do**

$z \leftarrow -F(x)$
for $j = 0, n - 1$ **do**
   $a \leftarrow \frac{\lambda_j}{\lambda_{j+1}}; \ b \leftarrow \lambda_j - 1$
   $z \leftarrow z + (as_{j+1} + bs_j)s_j^T z / \|s_j\|^2$
end for
$d \leftarrow (z - (1 - \lambda_n)s_n)/(1 + \lambda_n s_n^T z / \|s_n\|^2)$
Compute λ_{n+1} with a line search.
Terminate if the line search fails.
$s_{n+1} \leftarrow \lambda_{n+1}d; \ x \leftarrow x + s_{n+1}$
$n \leftarrow n + 1$
end while

As is the case with GMRES, the iteration can be restarted if there is no more room to store the vectors [30, 42]. Our MATLAB code `brsola.m` allows for this. A different approach, called **limited memory** in the optimization literature [54, 55], is to replace the oldest of the stored steps with the most recent one.

4.4 What Can Go Wrong?

Most of the problems you'll encounter are shared with the Newton–Krylov methods. When the nonlinear iteration converges slowly or the method completely fails, the preconditioner is one likely cause.

There are a few failure modes that are unique to Broyden's method, which, like the chord method, has no guarantee of global convergence.

4.4.1 Failure of the Line Search

There is no guarantee that a line search will succeed with Broyden's method. Our code `brsola.m` has a line search, but if you find that it fails, you may need to find a better preconditioner or switch to a Newton–Krylov method.

4.4.2 Failure to Converge

The local theory for Broyden states that the convergence is superlinear if the data x_0 and B_0 are good. If the data are poor or you use all available storage for updating B, the nonlinear iteration may fail. As with line search failure, better preconditioning may fix this.

4.5 Using `brsola.m`

`brsola.m` is an implementation of Broyden's method as described in Algorithm **broyden**. The user interface is similar to those of `nsold.m` and `nsoli.m`:

```
[sol, it_hist, ierr, x_hist] = brsola(x,f,tol, parms).
```

4.5.1   Input to `brsola.m`

The required data for `brsola.m` are x, the function f, and the tolerances for termination. The vector $tol = (\tau_a, \tau_r)$ contains the tolerances for the termination criterion (1.12).

   x and f must be column vectors of the same length. The syntax for f is

```
function = f(x).
```

   The *parms* array is

$$parms = [maxit, maxitl].$$

maxit is the upper limit on the nonlinear iterations; the default is 40. *maxitl* is the maximum number of nonlinear iterations before a restart (so $maxitl - 1$ vectors are stored for the nonlinear iteration). The default is 40.

4.5.2   Output from `brsola.m`

Exactly as in `nsoli.m`, the outputs are the solution *sol* and, optionally, a history of the iteration, an error flag, and the entire sequence $\{x_n\}$. We warn you again not to ask for the sequence $\{x_n\}$ unless you have the storage for this array. For large problems, asking for the iteration history $\{x_n\}$ by including *x_hist* in the argument list can expend all of MATLAB's storage. The code `heqmovie.m` in the directory for this chapter is an example of how to use `brsola.m` and the sequence of iterates to make a movie. The history array *it_hist* has three columns. The first is the Euclidean norm of the nonlinear residual $\|F(x)\|$, the second is the cumulative number of calls to F, and the third is the number of step-size reductions done in the line search.

   The error flag *ierr* is 0 if the nonlinear iteration terminates successfully. The failure modes are *ierr* = 1, which means that the termination criterion is not met after *maxit* iterations, and *ierr* = 2, which means that the step length was reduced 10 times in the line search without satisfaction of the sufficient decrease condition (1.21). Notice that we give the line search only 10 chances to satisfy (1.21), rather than the generous 20 given to `nsoli.m`. One can increase this by changing an internal parameter *maxarm* in the code.

4.6   Examples

Broyden's method, when working well, is superlinearly convergent in the terminal phase of the iteration. However, when the line search fails, Broyden's method is useless. For example, the line search will fail if you use `brsola.m` to solve the OZ equations from section 3.6.2 (unless you find a good preconditioner). Because of the uncertainty of the line search, Broyden's method is not as generally applicable as a Newton–Krylov method, where the Jacobian-vector product is highly accurate.

4.6.1   Chandrasekhar H-equation

We'll solve the same problem (equation, initial iterate, and tolerances) as we did in sections 2.7.3 and 3.6.1 with `brsola.m`. We compare `brsola.m` with both `nsoli.m` and `nsold.m` using the default choices of the parameters. The MATLAB code `heqbdemo.m` generated these results.

This fragment from `heqbdemo.m` is the call to `brsola.m`.

```
% Solve the H-equation with brsola.
x=ones(n,1);
[sol, it_hist, ierr] = brsola(x,'heq',tol);
```

`nsold` evaluates the Jacobian only once and takes 12 nonlinear iterations and 13 function evaluations to terminate. `nsoli` terminates in 5 iterations, but at a cost of 15 function evaluations. Since we can evaluate the Jacobian for the H-equation very efficiently, the overall cost is about the same. `broyden` is at its best for this kind of problem. We used the identity as the initial approximation for the Jacobian (i.e., we did not precondition); one can see that the nonlinear iteration is slower than the two Newton-based methods for the first few iterations, after which the updating takes effect. Broyden's method terminated after 7 nonlinear iterations and 8 function evaluations.

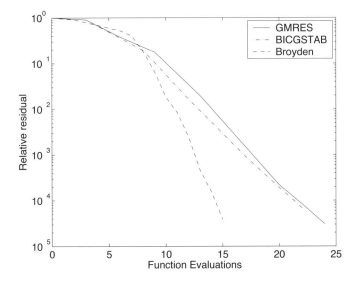

Figure 4.1. *Nonlinear residual versus calls to F. Right preconditioning.*

4.6.2   Convection-Diffusion Equation

In this section we compare Broyden's method to the right-preconditioned partial differential equation from section 3.6.3. The MATLAB code that generated this example is `pdebrr.m`. This is an interesting example because the line search in `brsola.m` succeeds, reducing the step length once on iterations 2 and 3. Contrast this with the two solves using `nsoli.m`, which required no reduction. In spite of the extra work in the line search, Broyden's method does best on this example.

Figure 4.1, one of the plots created by `pdebrr.m`, shows that simply counting iterations is not enough to compare the methods. While Broyden's method takes more nonlinear iterations, the cost in terms of calls to the function is significantly less.

For left preconditioning, the results (obtained with `pdebrl.m`) are similar. `nsoli.m` does not need the line search at all, but `brsola.m` reduces the step length once on the second nonlinear iteration. Even so, Broyden's method takes more than 20% fewer nonlinear function evaluations.

When storage is limited, Broyden's method is less impressive. In the case of the left-preconditioned convection-diffusion problem, for example, Broyden's method required 10 nonlinear iterations at a cost of 10 vectors of storage. Newton-GMRES, on the other hand, took at most 6 linear iterations for each nonlinear iteration.

4.7 Source Code for brsola.m

```matlab
1   function [sol, it_hist, ierr, x_hist] = brsola(x,f,tol, parms)
2   % BRSOLA   Broyden's Method solver, globally convergent
3   %          solver for f(x) = 0, Armijo rule, one vector storage
4   %
5   % C. T. Kelley, April 1, 2003
6   %
7   % This code comes with no guarantee or warranty of any kind.
8   %
9   % function [sol, it_hist, ierr, x_hist] = brsola.m(x,f,tol,parms)
10  %
11  % inputs:
12  %        initial iterate = x
13  %        function = f
14  %        tol = [atol, rtol] relative/absolute
15  %            error tolerances for the nonlinear iteration
16  %        parms = [maxit, maxdim]
17  %            maxit = maximum number of nonlinear iterations
18  %                default = 40
19  %            maxdim = maximum number of Broyden iterations
20  %                before restart, so maxdim-1 vectors are
21  %                stored
22  %                default = 40
23  %
24  % output:
25  %        sol = solution
26  %        it_hist(maxit,3) = 12 norms of nonlinear residuals
27  %            for the iteration, number function evaluations,
28  %            and number of steplength reductions
29  %        ierr = 0 upon successful termination
30  %        ierr = 1 if after maxit iterations
31  %            the termination criterion is not satisfied.
32  %        ierr = 2 failure in the line search. The iteration
33  %            is terminated if too many steplength reductions
34  %            are taken.
35  %
36  %    x_hist = matrix of the entire iteration history.
37  %            The columns are the nonlinear iterates. This
38  %            is useful for making movies, for example, but
39  %            can consume way too much storage. This is an
40  %            OPTIONAL argument. Storage is only allocated
41  %            if x_hist is in the output argument list.
42  %
43  % internal parameter:
44  %    debug = turns on/off iteration statistics display as
45  %            the iteration progresses
46  %
47  %    alpha = 1.d-4, parameter to measure sufficient decrease
48  %
49  %
50  %    maxarm = 10, maximum number of steplength reductions before
51  %            failure is reported
52  %
53  % set the debug parameter, 1 turns display on, otherwise off
54  %
55  debug=0;
56  %
57  % initialize it_hist, ierr, x_hist, and set the iteration parameters
58  %
59  ierr = 0; maxit=40; maxdim=39;
60  it_histx=zeros(maxit,3);
61  maxarm=10;
62  %
63  if nargin == 4
64      maxit=parms(1); maxdim=parms(2)-1;
65  end
66  if nargout==4
67      x_hist=x;
68  end
69  rtol=tol(2); atol=tol(1); n = length(x); fnrm=1; itc=0; nbroy=0;
70  %
71  % evaluate f at the initial iterate
72  % compute the stop tolerance
73  %
74  f0=feval(f,x);
75  fc=f0;
76  fnrm=norm(f0);
77  it_histx(itc+1,1)=fnrm; it_histx(itc+1,2)=0; it_histx(itc+1,3)=0;
78  fnrmo=1;
79  stop_tol=atol + rtol*fnrm;
80  outstat(itc+1, :) = [itc fnrm 0 0];
81  %
82  % terminate on entry?
83  %
84  if fnrm < stop_tol
85      sol=x;
86      return
87  end
88  %
89  % initialize the iteration history storage matrices
90  %
91  stp=zeros(n,maxdim);
92  stp_nrm=zeros(maxdim,1);
93  lam_rec=ones(maxdim,1);
94  %
95  % Set the initial step to -F, compute the step norm
96  %
97  lambda=1;
98  stp(:,1) = -fc;
99  stp_nrm(1)=stp(:,1)'*stp(:,1);
100
```

```
101  %
102  % main iteration loop
103  %
104  while(itc < maxit)
105  %
106      nbroy=nbroy+1;
107  %
108  % keep track of successive residual norms and
109  % the iteration counter (itc)
110  %
111      fnrmo=fnrm; itc=itc+1;
112  %
113  % compute the new point, test for termination before
114  % adding to iteration history
115  %
116      xold=x; lambda=1; iarm=0; lrat=.5; alpha=1.d-4;
117      x = x + stp(:,nbroy);
118      if nargout == 4
119          x_hist=[x_hist,x];
120      end
121      fc=feval(f,x);
122      fnrm=norm(fc);
123      ff0=fnrmo*fnrmo; ffc=fnrm*fnrm; lamc=lambda;
124  %
125  %
126  % Line search, we assume that the Broyden direction is an
127  % inexact Newton direction. If the line search fails to
128  % find sufficient decrease after maxarm steplength reductions
129  % brsola.m returns with failure.
130  %
131  % Three-point parabolic line search
132  %
133      while fnrm >= (1 - lambda*alpha)*fnrmo & iarm < maxarm
134          lambda=lambda*lrat;
135          if iarm==0
136              lambda=lambda*lrat;
137          else
138              lambda=parab3p(lamc, lamm, ff0, ffc, ffm);
139          end
140          lamm=lamc; ffm=ffc; lamc=lambda;
141          x = xold + lambda*stp(:,nbroy);
142          fc=feval(f,x);
143          fnrm=norm(fc);
144          ffc=fnrm*fnrm;
145          iarm=iarm+1;
146      end
147  %
148  % set error flag and return on failure of the line search
149  %
150      if iarm == maxarm
151          disp('line search failure in brsola.m ')
152          ierr=2;
153          it_hist=it_histx(1:itc+1,:);
154          sol=xold;
155          if nargout == 4
156              x_hist=[x_hist,x];
157          end
158          return;
159      end
160  %
161  % How many function evaluations did this iteration require?
162  %
163      it_histx(itc+1,1)=fnrm;
164      it_histx(itc+1,2)=it_histx(itc,2)+iarm+1;
165      if(itc == 1) it_histx(itc+1,2) = it_histx(itc+1,2)+1; end;
166      it_histx(itc+1,3)=iarm;
167  %
168  % terminate?
169  %
170      if fnrm < stop_tol
171          sol=x;
172          rat=fnrm/fnrmo;
173          outstat(itc+1, :) = [itc fnrm iarm rat];
174          it_hist=it_histx(1:itc+1,:);
175          it_hist(itc+1)=fnrm;
176          if debug==1
177              disp(outstat(itc+1,:))
178          end
179          return
180      end
181  %
182  % modify the step and step norm if needed to reflect the line
183  % search
184  %
185      lam_rec(nbroy)=lambda;
186      if lambda ~= 1
187          stp(:,nbroy)=lambda*stp(:,nbroy);
188          stp_nrm(nbroy)=lambda*lambda*stp_nrm(nbroy);
189      end
190  %
191  %
192      it_hist(itc+1)=fnrm;
193      rat=fnrm/fnrmo;
194      outstat(itc+1, :) = [itc fnrm iarm rat];
195      if debug==1
196          disp(outstat(itc+1,:))
197      end
198  %
199  %
200  % if there's room, compute the next search direction and step norm and
201  % add to the iteration history
202
```

```
203   %
204      if nbroy < maxdim+1
205         z=-fc;
206         if nbroy > 1
207            for kbr = 1:nbroy-1
208               ztmp=stp(:,kbr+1)/lam_rec(kbr+1);
209               ztmp=ztmp+(1 - 1/lam_rec(kbr))*stp(:,kbr);
210               ztmp=ztmp*lam_rec(kbr);
211               z=z+ztmp*((stp(:,kbr)'*z)/stp_nrm(kbr));
212            end
213         end
214   %
215   %        store the new search direction and its norm
216   %
217         a2=-lam_rec(nbroy)/stp_nrm(nbroy);
218         a1=1 - lam_rec(nbroy);
219         zz=stp(:,nbroy)'*z;
220         a3=a1*zz/stp_nrm(nbroy);
221         a4=1+a2*zz;
222         stp(:,nbroy+1)=(z-a3*stp(:,nbroy))/a4;
223         stp_nrm(nbroy+1)=stp(:,nbroy+1)'*stp(:,nbroy+1);
224   %
225   %
226   %
227      else
228   %
229   %        out of room, time to restart
230   %
231         stp(:,1)=-fc;
232         stp_nrm(1)=stp(:,1)'*stp(:,1);
233         nbroy=0;
234   %
235   %
236   %
237      end
238   %
239   % end while
240   end
241   %
242   % We're not supposed to be here, we've taken the maximum
243   % number of iterations and not terminated.
244   %
245   sol=x;
246   it_hist=it_histx(1:itc+1,:);
247   if nargout == 4
248      x_hist=[x_hist,x];
249   end
250   ier=1;
251   if debug==1
252      disp(outstat)
253   end
```

```
254   %
255   function lambdap = parab3p(lambdac, lambdam, ff0, ffc, ffm)
256   % Apply three-point safeguarded parabolic model for a line search.
257   %
258   % C. T. Kelley, April 1, 2003
259   %
260   % This code comes with no guarantee or warranty of any kind.
261   %
262   % function lambdap = parab3p(lambdac, lambdam, ff0, ffc, ffm)
263   %
264   % input:
265   %        lambdac = current steplength
266   %        lambdam = previous steplength
267   %        ff0 = value of \| F(x_c) \|^2
268   %        ffc = value of \| F(x_c + \lambdac d) \|^2
269   %        ffm = value of \| F(x_c + \lambdam d) \|^2
270   %
271   % output:
272   %        lambdap = new value of lambda given parabolic model
273   %
274   % internal parameters:
275   %        sigma0 = .1, sigma1 = .5, safeguarding bounds for the linesearch
276   %
277   %
278   %
279   % Set internal parameters.
280   %
281   sigma0 = .1; sigma1 = .5;
282   %
283   % Compute coefficients of interpolation polynomial.
284   %
285   % p(lambda) = ff0 + (c1 lambda + c2 lambda^2)/d1
286   %
287   % d1 = (lambdac - lambdam)*lambdac*lambdam < 0
288   %    so, if c2 > 0 we have negative curvature and default to
289   %        lambdap = sigam1 * lambda.
290   %
291   c2 = lambdam*(ffc-ff0)-lambdac*(ffm-ff0);
292   if c2 >= 0
293      lambdap = sigma1*lambdac; return
294   end
295   c1 = lambdac*lambdac*(ffm-ff0)-lambdam*lambdam*(ffc-ff0);
296   lambdap = -c1*.5/c2;
297   if lambdap < sigma0*lambdac, lambdap = sigma0*lambdac; end
298   if lambdap > sigma1*lambdac, lambdap = sigma1*lambdac; end
```

Bibliography

[1] E. ANDERSON, Z. BAI, C. BISCHOF, S. BLACKFORD, J. DEMMEL, J. DON-GARRA, J. DU CROZ, A. GREENBAUM, S. HAMMARLING, A. MCKENNEY, AND D. SORENSEN, *LAPACK Users Guide,* Third Edition, SIAM, Philadelphia, 1999.

[2] L. ARMIJO, *Minimization of functions having Lipschitz-continuous first partial derivatives,* Pacific J. Math., 16 (1966), pp. 1–3.

[3] U. M. ASCHER AND L. R. PETZOLD, *Computer Methods for Ordinary Differential Equations and Differential-Algebraic Equations,* SIAM, Philadelphia, 1998.

[4] K. E. ATKINSON, *An Introduction to Numerical Analysis,* Second Edition, John Wiley and Sons, New York, 1989.

[5] S. BALAY, W. D. GROPP, L. C. MCINNES, AND B. F. SMITH, *Portable, Extensible Toolkit for Scientific Computation (PETSc) home page.* http://www.mcs.anl.gov/petsc.

[6] S. BALAY, W. D. GROPP, L. C. MCINNES, AND B. F. SMITH, *PETSc 2.0 Users Manual,* Tech. Rep. ANL-95/11 - Revision 2.0.28, Argonne National Laboratory, Argonne, IL, 2000.

[7] M. J. BOOTH, A. G. SCHLIPER, L. E. SCALES, AND A. D. J. HAYMET, *Efficient solution of liquid state integral equations using the Newton-GMRES algorithm,* Comput. Phys. Comm., 119 (1999), pp. 122–134.

[8] K. E. BRENAN, S. L. CAMPBELL, AND L. R. PETZOLD, *Numerical Solution of Initial-Value Problems in Differential-Algebraic Equations,* Vol. 14 in Classics in Applied Mathematics, SIAM, Philadelphia, 1996.

[9] W. L. BRIGGS, V. E. HENSON, AND S. F. MCCORMICK, *A Multigrid Tutorial,* Second Edition, SIAM, Philadelphia, 2000.

[10] P. N. BROWN AND A. C. HINDMARSH, *Matrix-free methods for stiff systems of ODE's,* SIAM J. Numer. Anal., 23 (1986), pp. 610–638.

[11] P. N. BROWN AND A. C. HINDMARSH, *Reduced storage matrix methods in stiff ODE systems,* J. Appl. Math. Comput., 31 (1989), pp. 40–91.

[12] P. N. BROWN, A. C. HINDMARSH, AND L. R. PETZOLD, *Using Krylov methods in the solution of large-scale differential-algebraic systems*, SIAM J. Sci. Comput., 15 (1994), pp. 1467–1488.

[13] P. N. BROWN AND Y. SAAD, *Hybrid Krylov methods for nonlinear systems of equations*, SIAM J. Sci. Stat. Comput., 11 (1990), pp. 450–481.

[14] C. G. BROYDEN, *A class of methods for solving nonlinear simultaneous equations*, Math. Comput., 19 (1965), pp. 577–593.

[15] I. W. BUSBRIDGE, *The Mathematics of Radiative Transfer*, no. 50 in Cambridge Tracts, Cambridge University Press, Cambridge, U.K., 1960.

[16] S. L. CAMPBELL, I. C. F. IPSEN, C. T. KELLEY, AND C. D. MEYER, *GMRES and the minimal polynomial*, BIT, 36 (1996), pp. 664–675.

[17] S. CHANDRASEKHAR, *Radiative Transfer*, Dover, New York, 1960.

[18] Z.-M. CHEN AND B. M. PETTITT, *Non-isotropic solution of an OZ equation: Matrix methods for integral equations*, Comput. Phys. Comm., 85 (1995), pp. 239–250.

[19] C. T. KELLEY, D. E. KEYES, AND T. COFFEY, *Pseudo-transient continuation and differential-algebraic equations*, SIAM J. Sci. Comput., to appear.

[20] T. F. COLEMAN AND J. J. MORÉ, *Estimation of sparse Jacobian matrices and graph coloring problems*, SIAM J. Numer. Anal., 20 (1983), pp. 187–209.

[21] A. R. CURTIS, M. J. D. POWELL, AND J. K. REID, *On the estimation of sparse Jacobian matrices*, J. Inst. Math. Appl., 13 (1974), pp. 117–119.

[22] R. S. DEMBO, S. C. EISENSTAT, AND T. STEIHAUG, *Inexact Newton methods*, SIAM J. Numer. Anal., 19 (1982), pp. 400–408.

[23] J. W. DEMMEL, *Applied Numerical Linear Algebra*, SIAM, Philadelphia, 1997.

[24] J. E. DENNIS, JR., AND R. B. SCHNABEL, *Numerical Methods for Unconstrained Optimization and Nonlinear Equations*, Vol. 16 in Classics in Applied Mathematics, SIAM, Philadelphia, 1996.

[25] P. DEUFLHARD, *Adaptive Pseudo-Transient Continuation for Nonlinear Steady State Problems*, Tech. Rep. 02-14, Konrad-Zuse-Zentrum für Informationstechnik, Berlin, March 2002.

[26] P. DEUFLHARD, R. W. FREUND, AND A. WALTER, *Fast secant methods for the iterative solution of large nonsymmetric linear systems*, Impact Comput. Sci. Engrg., 2 (1990), pp. 244–276.

[27] J. J. DONGARRA, J. BUNCH, C. B. MOLER, AND G. W. STEWART, *LINPACK Users' Guide*, SIAM, Philadelphia, 1979.

[28] S. C. EISENSTAT AND H. F. WALKER, *Globally convergent inexact Newton methods*, SIAM J. Optim., 4 (1994), pp. 393–422.

[29] S. C. EISENSTAT AND H. F. WALKER, *Choosing the forcing terms in an inexact Newton method*, SIAM J. Sci. Comput., 17 (1996), pp. 16–32.

[30] M. ENGELMAN, G. STRANG, AND K. J. BATHE, *The application of quasi-Newton methods in fluid mechanics*, Internat. J. Numer. Methods Engrg., 17 (1981), pp. 707–718.

[31] R. W. FREUND, *A transpose-free quasi-minimal residual algorithm for non-Hermitian linear systems*, SIAM J. Sci. Comput., 14 (1993), pp. 470–482.

[32] G. H. GOLUB AND C. F. VAN LOAN, *Matrix Computations*, Third Edition, Johns Hopkins Studies in the Mathematical Sciences, Johns Hopkins University Press, Baltimore, 1996.

[33] A. GREENBAUM, *Iterative Methods for Solving Linear Systems*, Vol. 17 in Frontiers in Applied Mathematics, SIAM, Philadelphia, 1997.

[34] A. GRIEWANK, *Evaluating Derivatives: Principles and Techniques of Algorithmic Differentiation*, Vol. 19 in Frontiers in Applied Mathematics, SIAM, Philadelphia, 2000.

[35] M. R. HESTENES AND E. STEIFEL, *Methods of conjugate gradient for solving linear systems*, J. Res. Nat. Bureau Standards, 49 (1952), pp. 409–436.

[36] D. J. HIGHAM, *Trust region algorithms and timestep selection*, SIAM J. Numer. Anal., 37 (1999), pp. 194–210.

[37] D. J. HIGHAM AND N. J. HIGHAM, *MATLAB Guide*, SIAM, Philadelphia, 2000.

[38] P. HOVLAND AND B. NORRIS, *Argonne National Laboratory Computational Differentiation Project*, Argonne National Laboratory, Argonne, IL, 2002. http://www-fp.mcs.anl.gov/autodiff/

[39] *IEEE Standard for Binary Floating Point Arithmetic, Std* 754-1885, IEEE, Piscataway, NJ, 1985.

[40] H. B. KELLER, *Numerical Solution of Two Point Boundary Value Problems*, Vol. 24 in CBMS-NSF Regional Conference Series in Applied Mathematics, SIAM, Philadelphia, 1976.

[41] C. T. KELLEY, *Solution of the Chandrasekhar H-equation by Newton's method*, J. Math. Phys., 21 (1980), pp. 1625–1628.

[42] C. T. KELLEY, *Iterative Methods for Solving Linear and Nonlinear Equations*, Vol. 16 in Frontiers in Applied Mathematics, SIAM, Philadelphia, 1995.

[43] C. T. KELLEY, *Iterative Methods for Optimization*, Vol. 18 in Frontiers in Applied Mathematics, SIAM, Philadelphia, 1999.

[44] C. T. KELLEY AND D. E. KEYES, *Convergence analysis of pseudo-transient continuation*, SIAM J. Numer. Anal., 35 (1998), pp. 508–523.

[45] C. T. KELLEY, C. T. MILLER, AND M. D. TOCCI, *Termination of Newton/chord iterations and the method of lines*, SIAM J. Sci. Comput., 19 (1998), pp. 280–290.

[46] C. T. KELLEY AND T. W. MULLIKIN, *Solution by iteration of H-equations in multigroup neutron transport*, J. Math. Phys., 19 (1978), pp. 500–501.

[47] C. T. KELLEY AND B. M. PETTITT, *A Fast Algorithm for the Ornstein–Zernike Equations*, Tech. Rep. CRSC-TR02-12, North Carolina State University, Center for Research in Scientific Computation, Raleigh, NC, April 2002.

[48] T. KERKHOVEN AND Y. SAAD, *On acceleration methods for coupled nonlinear elliptic systems*, Numer. Math., 60 (1992), pp. 525–548.

[49] J. N. LYNESS AND C. B. MOLER, *Numerical differentiation of analytic functions*, SIAM J. Numer. Anal., 4 (1967), pp. 202–210.

[50] T. A. MANTEUFFEL AND S. V. PARTER, *Preconditioning and boundary conditions*, SIAM J. Numer. Anal., 27 (1990), pp. 656–694.

[51] J. J. MORÉ, B. S. GARBOW, AND K. E. HILLSTROM, *User Guide for MINPACK-1*, Tech. Rep. ANL-80-74, Argonne National Laboratory, Argonne, IL, 1980.

[52] T. W. MULLIKIN, *Some probability distributions for neutron transport in a half space*, J. Appl. Prob., 5 (1968), pp. 357–374.

[53] N. M. NACHTIGAL, S. C. REDDY, AND L. N. TREFETHEN, *How fast are nonsymmetric matrix iterations?*, SIAM J. Matrix Anal. Appl., 13 (1992), pp. 778–795.

[54] J. L. NAZARETH, *Conjugate gradient methods less dependent on conjugacy*, SIAM Rev., 28 (1986), pp. 501–512.

[55] J. NOCEDAL, *Theory of algorithms for unconstrained optimization*, Acta Numer., 1 (1991), pp. 199–242.

[56] L. S. ORNSTEIN AND F. ZERNIKE, *Accidental deviations of density and opalescence at the critical point of a single substance*, Proc. Konink. Nederl. Akad. Wetensch., 17 (1914), pp. 793–806.

[57] J. M. ORTEGA AND W. C. RHEINBOLDT, *Iterative Solution of Nonlinear Equations in Several Variables*, Academic Press, New York, 1970.

[58] M. L. OVERTON, *Numerical Computing with IEEE Floating Point Arithmetic*, SIAM, Philadelphia, 2001.

[59] M. PERNICE AND H. F. WALKER, *NITSOL: A Newton iterative solver for nonlinear systems*, SIAM J. Sci. Comput., 19 (1998), pp. 302–318.

[60] M. J. D. POWELL, *A hybrid method for nonlinear equations*, in Numerical Methods for Nonlinear Algebraic Equations, P. Rabinowitz, ed., Gordon and Breach, New York, 1970, pp. 87–114.

[61] K. RADHAKRISHNAN AND A. C. HINDMARSH, *Description and Use of LSODE, the Livermore Solver for Ordinary Differential Equations*, Tech. Rep. URCL-ID-113855, Lawrence Livermore National Laboratory, Livermore, CA, December 1993.

[62] Y. SAAD, *ILUM: A multi-elimination ILU preconditioner for general sparse matrices*, SIAM J. Sci. Comput., 17 (1996), pp. 830–847.

[63] Y. SAAD, *Iterative Methods for Sparse Linear Systems*, Second Edition, SIAM, Philadelphia, 2003.

[64] Y. SAAD AND M. H. SCHULTZ, *GMRES: A generalized minimal residual algorithm for solving nonsymmetric linear systems*, SIAM J. Sci. Stat. Comput., 7 (1986), pp. 856–869.

[65] R. B. SCHNABEL, J. E. KOONTZ, AND B. E. WEISS, *A modular system of algorithms for unconstrained minimization*, ACM TOMS, 11 (1985), pp. 419–440. ftp://ftp.cs.colorado.edu/users/uncmin/tape.jan30/shar

[66] V. E. SHAMANSKII, *A modification of Newton's method*, Ukrain. Mat. Zh., 19 (1967), pp. 133–138 (in Russian).

[67] L. F. SHAMPINE, *Implementation of implicit formulas for the solution of ODEs*, SIAM J. Sci. Stat. Comput., 1 (1980), pp. 103–118.

[68] L. F. SHAMPINE AND M. W. REICHELT, *The MATLAB ODE suite*, SIAM J. Sci. Comput., 18 (1997), pp. 1–22.

[69] J. SHERMAN AND W. J. MORRISON, *Adjustment of an inverse matrix corresponding to changes in the elements of a given column or a given row of the original matrix (abstract)*, Ann. Math. Stat., 20 (1949), p. 621.

[70] J. SHERMAN AND W. J. MORRISON, *Adjustment of an inverse matrix corresponding to a change in one element of a given matrix*, Ann. Math. Stat., 21 (1950), pp. 124–127.

[71] K. SIGMON AND T. A. DAVIS, *MATLAB Primer,* Sixth Edition, CRC Press, Boca Raton, FL, 2002.

[72] B. SMITH, P. BJØRSTAD, AND W. GROPP, *Domain Decomposition: Parallel Multilevel Methods for Elliptic Partial Differential Equations*, Cambridge University Press, Cambridge, U.K., 1996.

[73] W. SQUIRE AND G. TRAPP, *Using complex variables to estimate derivatives of real functions*, SIAM Rev., 40 (1998), pp. 110–112.

[74] G. W. STEWART, *Introduction to Matrix Computations*, Academic Press, New York, 1973.

[75] A. G. TAYLOR AND A. C. HINDMARSH, *User Documentation for KINSOL, a Nonlinear Solver for Sequential and Parallel Computers*, Tech. Rep. UCRL-ID-131185, Lawrence Livermore National Laboratory, Center for Applied Scientific Computing, Livermore, CA, July 1998.

[76] L. N. TREFETHEN AND D. BAU III, *Numerical Linear Algebra*, SIAM, Philadelphia, 1997.

[77] H. A. VAN DER VORST, *BI-CGSTAB: A fast and smoothly converging variant of BI-CG for the solution of nonsymmetric linear systems*, SIAM J. Sci. Stat. Comput., 13 (1992), pp. 631–644.

[78] L. T. WATSON, S. C. BILLUPS, AND A. P. MORGAN, *Algorithm 652: HOMPACK: A suite of codes for globally convergent homotopy algorithms*, ACM Trans. Math. Software, 13 (1987), pp. 281–310.

Index